George Harley

The Urine and its Derangements

With the Application of Physiological Chemistry to the Diagnosis and Treatment of

Constitutional

George Harley

The Urine and its Derangements
With the Application of Physiological Chemistry to the Diagnosis and Treatment of Constitutional

ISBN/EAN: 9783337140212

Printed in Europe, USA, Canada, Australia, Japan

Cover: Foto ©berggeist007 / pixelio.de

More available books at **www.hansebooks.com**

THE URINE

AND ITS DERANGEMENTS,

APPLICATION OF PHYSIOLOGICAL CHEMISTRY

TO THE

DIAGNOSIS AND TREATMENT

OF

Constitutional, as well as Local Diseases:

BEING A COURSE OF ORIGINAL LECTURES

DELIVERED AT UNIVERSITY COLLEGE, LONDON,

BY

GEORGE HARLEY, M.D., F.R.S.,

FELLOW OF THE ROYAL COLLEGE OF PHYSICIANS; CORRESPONDING MEMBER OF THE ACADEMY OF SCIENCES OF BAVARIA, THE ACADEMY OF MEDICINE OF MADRID, OF THE MEDICAL SOCIETIES OF WURBBURG, HALLE, BADEN, ETC.; FORMERLY PROFESSOR IN UNIVERSITY COLLEGE, AND PHYSICIAN TO UNIVERSITY COLLEGE HOSPITAL.

LONDON:
J. AND A. CHURCHILL, NEW BURLINGTON STREET.

1872.

PREFACE.

THE lectures composing this book were originally delivered to a mixed class of medical men, and advanced students, over a period of ten years, and afterwards published in detached portions, and at varying intervals, in the *Medical Times and Gazette*. Two of them only (Diabetes and Albuminuria) were subsequently reproduced separately.

The favourable manner in which the course was received, both by the gentlemen who honoured the class-room with their presence, and by the profession generally, when it had assumed the printed form, has induced the author to think that as he has now ceased giving oral instruction, the re-publication of the revised and enlarged lectures, in a single volume, may not be entirely unacceptable to the younger members of the profession; seeing that as yet the facilities for the study of Physiological and Pathological Chemistry, as applied to the Diagnosis and Treatment of Constitutional as well as Local Diseases, are but very imperfectly provided for them in the majority of our medical schools—a circumstance which every enlightened practitioner of medicine cannot fail to deplore; for, as was aptly stated in one of the annual editorial addresses to students (a)— "Chemistry has become a very important hand-maid to medicine, since there are few forms of disease to the diagnosis and treatment of which it may not in some way be made subservient; whilst for the due investigation of others, as the entire class of urinary affections, it is absolutely indispensable."

25, HARLEY STREET, CAVENDISH SQUARE, W.

(a) *Lancet*, vol. ii., p. 335, Students' Number for 1863.

CONTENTS.

THE URINE AND ITS DERANGEMENTS

AS AN AID TO THE

DIAGNOSIS AND TREATMENT OF

CONSTITUTIONAL, AS WELL AS LOCAL DISEASES.

———◆———

In beginning a course of lectures on the urine, intended as an aid to the diagnosis and treatment of disease, the first question requiring an immediate answer is :—

WHAT IS URINE?

Urine is nothing more nor less than a collection of the liquid and solid effete products of the frame, and its composition fluctuates in exact proportion as the processes of life fluctuate.

Consequently, it differs very materially in different species of animals, and even varies with the varying conditions of the same animal.

The state of the urine has, with justice, long been regarded as a key to the condition of the body; and hence we find that the Physicians of antiquity, like those of modern times, sought in the composition of the urine of their patients, a clue to the nature of the diseases under which they laboured, an index to their prognosis, and a guide to their successful treatment.

It is true that uroscopic charlatanism has, more or less, in all ages, laid human credulity under contribution, and thereby brought discredit upon, and retarded the advance of legitimate

B

science; but this must not deter the honest Practitioner from reaping the advantages which a knowledge of this secretion affords.

An examination of the urine not only enables us to arrive at a correct diagnosis of diseases attacking the urinary organs themselves, but in many cases yields most important information regarding the nature of morbid changes occurring in other and distant parts of the frame. One must not, however, be led away with the idea that the urine is the royal road to knowledge, or he will be doomed to disappointment; nor must we flatter ourselves that its study is an easy one, requiring neither previous preparation nor subsequent application; for on the contrary, in order to make urology bear fruit, we require to employ in its cultivation both perseverance and skill.

It is not a little remarkable that this secretion, which inspires most of us with a feeling of aversion, should not only have rendered incalculable service to clinical Medicine, and yielded brilliant discoveries to chemistry, but also solved many of the most abstruse problems in physiology. Notwithstanding all this, it has not done nearly as much as it might have done, nor as it ought to have done; but this is owing to no defect on its side, but in consequence of its votaries having always been too eager to profit by its pathology ere they had mastered the first rudiments of its physiology.

In fact, until within the last year or two, men have been constantly attempting to place, as it were, the pyramid of Medical science on its apex instead of on its base, and the result has naturally been that they reaped little benefit from their labours. It seems, indeed, as if they were only now becoming alive to the all-important fact, that physiology is the only true basis of rational Medicine, and that in direct proportion as they adopt its principles will be the success of their labours in advancing the healing art. Such being the views now held by the pioneers of clinical Medicine in all countries, you will readily understand why, before entering on the pathology of the urine, I first try· to give you a succinct account of its physiology.

It has been said that urine is nothing more than a collection of the effete products of the frame, and, consequently, that ⋅it differs in different species of animals. If we cast our eyes over the whole animal economy, we shall discover that all urines are not necessarily liquid ; on the contrary, we shall find that there are many species of animals that pass solid urine ; and thus at the very threshold of our inquiry we perceive this secretion naturally dividing itself into two great classes—the *solid* and the *liquid* urines.

Solid Urine.

In all animals devoid of a urinary bladder, and in which the ureters open into the rectum, the urine is solid. Thus, for example, the urine of serpents is passed in a compact mass, varying, with the size of the animal, from that of a pea to that of an orange. Yet, notwithstanding the peculiar appearance of this specimen of urine from the boa constrictor, it differs from liquid urine in only one particular —the absence of water. By the simple addition of distilled water to it I can produce as perfect a urine as that of the human being ; for, solid though it be, it contains all the urinary ingredients,—urea, uric acid, phosphates, etc. Here, for example, are two spatulas, on one of which I place a fragment of serpent's urine, and on the other, some of the solids from evaporated human urine ; to each is added a couple of drops of strong nitric acid, and you observe that both effervesce. I now heat them over the flame of the spirit-lamp in order to drive away the excess of acid, and to their yellow-coloured residues add a drop of strong liquor ammonia, when instantly, you perceive, both assume a magnificent purple tint. This colour is due to the presence of purpurate of ammonia, produced by the action of the reagents on the uric acid contained in the urines. You observe, too, that the urine of the snake —for it has become much more crimson than the other— appears to be the richest in that material. The uric acid of commerce is indeed almost entirely obtained from serpent's urine; consequently, these excrementitious masses sometimes cost as much as thirty shillings per pound.

Hitherto in these lectures I have generally said that all reptiles have solid urines; but, as this has occasionally led my hearers into a mistake, from the fact that, in common language, frogs and toads are reptiles, while, scientifically speaking, they do not belong to the class *Reptilia*, I must mention that these animals have distinct urinary bladders, and possess liquid urines. Frog's urine, for example, is a clear, transparent liquid, containing urea, phosphate of lime, chloride of sodium, and other urinary ingredients, just like human urine. The true reptile, on the other hand, has always a solid urine. Thus it is that the chameleon, which, like the serpent, is a true reptile, passes excrements containing urates, oxalates, phosphates, and, according to Kletzinsky, even xanthic oxide,—one of the rare urinary substances. Do not for a moment imagine, however, that solid urine is one of the characteristics of the lower animal. On the contrary, we meet with it even in the higher classes. Birds have solid urines. Guano—the excrement of the sea-fowl—is in great part urine, and besides the principle, guanine, contains urate of ammonia, oxalate of lime, and ammoniaco-magnesian phosphates.

Again, we find solid urines throughout the whole insect tribe. The red excrements of the butterfly contain both urates, phosphates, and oxalates; and so it is with the excrements of all other species of insects. We see, then, as was before said, that the only difference between the solid and the liquid urines is in the absence of water.

Liquid Urine.

Having made these remarks on solid, we shall now inquire into the nature of the liquid urines. The liquid urines are found throughout the whole class *Mammalia*, and present three such well-marked varieties, both as regards physical appearance and chemical composition, that it may be said they naturally divide themselves, according to the species of animal, into the three separate groups of carnivorous, herbivorous, and omnivorous urines.

The urine of the first class (carnivora) is characterised as being a clear, transparent, light-coloured liquid, possessing an acid reaction, and rarely depositing anything on cooling.

The urine of the second class (herbivora), on the other hand, is recognised as being a dark-coloured liquid, with a strongly alkaline reaction, and depositing a copious sedimen within twenty-four hours after being passed.

The urine of the third class (omnivora) lies, as it were, between the two. It is clear, slightly acid, somewhat darker in colour than that of the carnivora, but considerably paler than that of the herbivora, and only occasionally deposits a sediment on cooling.

In chemical constitution these urines present as marked features of difference as they do in physical appearance. Thus, for example, while uric acid (Fig. 1), a beautiful crystalline body, exists in the urine of the carnivora, it is entirely wanting in that of the herbivora.

The urine of the herbivora, on the other hand, contains, in its stead, a quantity of a totally distinct organic acid, namely. hippuric, which not only differs in chemical composition, but also in crystalline form (Fig. 2). The urine of the omnivora is again found to stand between the two, and contains a portion of uric as well as hippuric acid.

Even in the nature of their inorganic ingredients, these urines differ materially; for, while both alkaline and earthy

FIG. 1.

FIG. 1.—Uric acid from human urine, magnified 250 diameters.

phosphates are abundant in the urine of the carnivora, they are entirely absent from that of the herbivora. — their place being there supplied by alkaline and earthy carbonates. Here the urine of the omnivora still occupies the medium position, and claims a share of both.

Having now seen that the urine varies in the different species of animals, we are in a measure prepared to consider how and why it should vary at different times in the same animal. The variations both as regards quantity and quality we shall find to depend upon the state of the body, the kind of food and drink, the amount of exercise, the climate, and an infinite number of minor causes, the influence of which will afterwards appear.

FIG. 2.

FIG. 2.—Hippuric acid from human urine, magnified 80 diameters.

As it is with the diseases of man alone that we have to do, we must now direct our attention specially to

HUMAN URINE.

The most readily observable properties of the human urine are—its colour, smell, taste, reaction, and specific gravity.

The Colour of the Human Urine.—The colour may be said to be quite characteristic, as no other animal liquid possesses a similar one. In health, it fluctuates between a pale straw and a brownish-yellow tint. It is due to the existence of a peculiar pigment—urohæmatin—and upon the relative proportion of this colouring matter and the water do all the varying tints of the healthy urine depend. Consequently, you now observe that I artificially produce from this intensely-coloured urine each of the different normal shades by the simple addition of pure water. Hence, we find that, so long as the body remains in health, the larger the quantity of urine daily excreted, the paler the tint; the less the amount, the higher the colour.

Although, as I shall afterwards have occasion to show you, normally-coloured urine does not necessarily exclude the possibility of disease, yet it is in general the presence of orange, red, brown, black, green, or blue urine that indicates the existence of the graver maladies. Still, even the presence of high-coloured urine, when an average amount is passed, is a sign which ought never to be lost sight of, as it invariably indicates the existence of a pathological condition of system. The import of this, like that of the other tints, will afterwards occupy our attention; meanwhile we must pass on to the consideration of the next physical condition of normal urine, namely :—

The Smell of the Urine.—Immediately after being passed from the bladder human urine has a mawkish aromatic odour. The smell of the urine is in general quite characteristic of the species of animal. All of you are acquainted with the smell of cat's urine, and I dare say most of you are equally familiar with the peculiar odour of the urine of the horse and cow. The urine in general smells like the fat of the animal; sheep's urine, for example, smells exactly like mutton suet. Many foods and drinks transmit to the urine their peculiar odour. To such an extent, too, may this be the case, that one is occasionally able, from the smell of the urine alone, to tell the kind of solid or liquid of which the patient has partaken. Thus, turpentine, copaiba, garlic, and many other aromatic substances, may be recognised in the urine. All of you must have observed what a strong odour asparagus imparts to the renal secretion. It not unfrequently happens that the mere breathing of odoriferous substances communicates their characteristic aroma to the urine; and the more delicate the individual, the more marked is this effect. The abnormal odours of the urine, such as we meet with in jaundice, Bright's disease, diabetes, etc., will each occupy our attention in its appropriate place.

The Taste of the Urine.—The taste of normal urine is saltish bitter, but the flavour varies very much in disease, being sometimes as sweet as honey—in diabetes; and at other times as bitter as gall—in icterus. In cases of urgent necessity, where the urine has been used as drink, it is said greatly to increase the thirst. Holwell, when confined in the black-hole at Calcutta, found the thirst, which proved so fatal to many, relieved by sucking his perspiration, but not by drinking his urine.

The Reaction of Human Urine.—When freshly passed the urine is slightly acid, the acidity becoming more marked during the first hour or two. According to Dr. Bence Jones it is neutral, or even alkaline, immediately after taking food, again gradually becoming more and more acid up to the time of the next meal. Although I have been unable to verify this statement on perfectly healthy individuals, I see nothing improbable in it, if the person experimented upon has partaken largely of vegetable food. The urine of dogs fed on flesh is acid, but if the animals are fed for several days on vegetable diet it becomes alkaline. The urine of rabbits, on the other hand, is normally alkaline; but if animal food be forced upon them, or better still, if for a time they receive no food at all, and are consequently obliged to live upon their own tissues, the urine becomes acid, and at the same time assumes many of the other characteristics of the carnivorous animal. It is, therefore, highly probable that diet will, in a similar way, influence the reaction of the human urine. It may be laid

down as a rule, however, that if the urine of a person living
on mixed diet becomes alkaline in less than twenty-four hours
after being passed, there exists some disease, either of the
general system or of the urinary organs, which demands immediate attention.

It must not be forgotten that many medicines—alkaline
carbonates, for example—as well as alkaline citrates, tartrates,
and acetates, which are converted, in their passage through
the body, into carbonates, cause even the urine of healthy
individuals to become for a time alkaline.

The acidity of normal urine depends on the united presence
of acid phosphate of soda, uric, (a) hippuric, and lactic acids.
Neither the hydrochloric nor sulphuric acids, which, as will
afterwards be seen, are always present in normal urine, have any
share in producing the acidity, for they are both combined with
bases in the form of neutral salts.   After standing for a greater
or lesser time, all urines become alkaline.   This arises from
the urea being decomposed into the carbonate of ammonia, in
consequence of a putrefactive fermentation being induced by
the presence of such animal matters as mucus from the
bladder and urinary passages.   The carbonate of ammonia is
a strongly alkaline salt; so alkaline, indeed, that putrid urine
was at one time in great request for the purpose of removing
grease from cloth.   This property the urine entirely owes to
the presence of the carbonate of ammonia, which unites with
the fatty matters of the cloth, and forms an alkaline soap,
readily removable by washing.   The peculiar odour, so
characteristic of putrid urine, is also chiefly due to the
carbonate of ammonia derived from the decomposed urea.

It occasionally happens that this decomposition occurs

(a) Some have thought that all the uric acid is united with part of the
soda, of what would otherwise have been the neutral phosphate of soda ;
but this is an error, as is seen by the fact, that in the most naturally acid
urines the uric acid is invariably found in a free state, as is proved by its
spontaneously crystallising when the urine cools.

within the frame ; and hence in certain diseased states of the bladder and urinary passages the urine is alkaline at the moment of being voided, while in certain others of the general system it becomes so very shortly after being passed.

The reaction of the urine is most readily ascertained with litmus paper. Some recommend neutral paper for this purpose, as one kind does for both acid and alkaline urine; but, as you now see, the change produced in the neutral paper is very slight, whereas with the more intensely coloured papers, one of which, as you perceive, is of a distinct blue, while the other is as evidently red, no difficulty can possibly arise. The blue paper I dip into this dog's urine, the acidity of which instantly turns it red; while the red paper I dip into the alkaline urine of the horse, and, as you observe, it is as rapidly changed to blue. Soft papers are usually more sensitive than hard.

When it is deemed desirable to ascertain the exact amount of free acid daily excreted, a standard solution of caustic soda is employed of such strength, that one cubic centimetre neutralises ten milligrammes of dry oxalic acid. This solution is gradually added to a given quantity of urine until the acid reaction disappears, and the calculation for the whole quantity passed in the twenty-four hours made accordingly. (a) It has been ascertained that healthy adults daily pass, on an average, an amount of free acid equal to about 1·7 grammes (27 grains) of dry oxalic acid.

This mode of estimating the amount of free acid present in urine is unfortunately too troublesome to admit of its being generally employed at the bedside, so that we are forced to guess at the acidity of the secretion from the intensity of the stain it produces on litmus paper.

It occasionally happens that the urine is recognised as being abnormally acid from the presence of brownish-red crystals of

---

(a) The mode of preparing and employing this solution will occupy our attention on a future occasion.

uric acid on the bottom and sides of the chamber-vessel; for free uric acid never crystallises spontaneously, except in cases where the urine contains an excess of acid, as, for example, in febrile and inflammatory affections, especially those in which the liver, heart, or lungs are implicated.

THE SPECIFIC GRAVITY OF HUMAN URINE.

The specific gravity of healthy urine ranges between 1002 (after drinking) and 1030 (after food), pure water being 1000. It is never lighter than water. In disease its density may be as high as 1060. The variations in the specific gravity of the healthy secretion depend upon the quantity of solids it contains, and that again varies with the time of the day, the constitution of the individual, the food and drink, and the amount of exercise taken. The urine that is passed immediately after drinking, and hence called " urina potûs," is pale-coloured, faintly acid, and of low specific gravity (1002 to 1015). That passed in the morning, after sleep, and named "urina sanguinis," from being supposed to come directly from the blood, is darker in colour, of a more acid reaction, and a higher specific gravity (1015 to 1020); while that voided some hours after food, and named, accordingly, " urina cibi," although having a still higher specific gravity than the last (1020 to 1030), is neither so dark in colour nor so acid in reaction as the morning's urine.

Seeing that the different urines voided at different times throughout the day vary so much in specific gravity, it is easily understood how dangerous it must be to draw any conclusions from observations made on only one of them, and proves the necessity of collecting the whole urine passed during the twenty-four hours, and estimating the specific gravity therefrom. When such is done, it is found that the average specific gravity varies in health from 1015 to 1025;

pregnancy being the only exception to this rule. In it, and more especially during the latter months, the average specific gravity may reach as high as 1033, and yet the individual be in the enjoyment of the most perfect health. I have even noticed that of the urina sanguinis in such cases as much as 1030. Were such changes as these observed in a non-pregnant female, they would infallibly point to an abnormal condition of system, such as the existence of febrile or inflammatory disease. In pneumonia and pleurisy the specific gravity of the urine is often as high as 1035, and in fever it has even been noticed to reach 1038. On the other hand, when the average specific gravity is abnormally low (less than 1015), we may with equal safety suspect the presence of some exhausting non-inflammatory complaint, such, for example, as Bright's disease, in which it may be as low as 1008, or as cholera, where it has been known to fall even to 1006. It may be stated as a rule, that the lower the specific gravity in these affections, the more unfavourable is the prognosis; the higher the specific gravity, the less dangerous the case. (a) The specific gravity of urine is estimated in three different ways. First, by means of density beads (Fig. 3, A), such as were formerly used in ascertaining the specific gravity of spirit.

These beads are gradually dropped into the urine until one sinks to the bottom of the vessel, when, by reading off the number on the last bead that floated, the specific gravity of the liquid is ascertained. In the urine glass (B), bead 20, being too heavy, has sunk to the bottom, while bead 15 is floating at the top. The specific gravity of the liquid is, therefore, 1015. This method is open to two objections :—1st. It does not come nearer than five degrees, for the beads are numbered 5, 10, 15, 20, 25, etc. ; and 2nd. It is exceedingly tedious in its mode of employment.

(a) The exceptions to this rule will afterwards be pointed out.

The second method of ascertaining the specific gravity is by what is called a urinometer (Fig. 3, c)—a most convenient

FIG. 3.

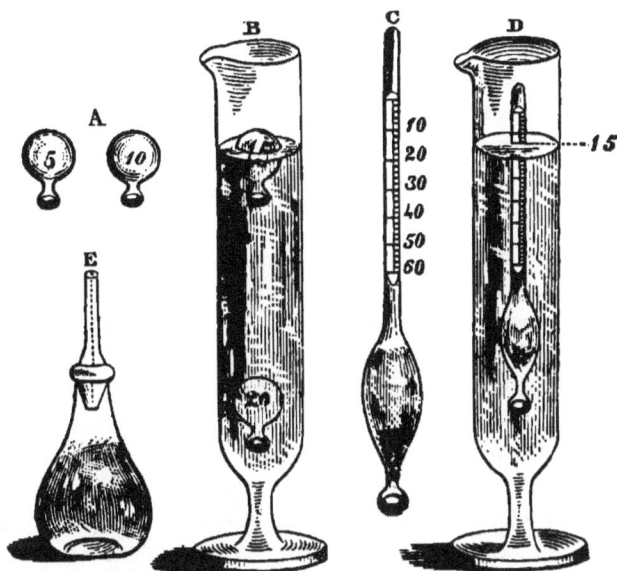

little instrument, consisting of a blown glass float, a small bulb weighted with mercury, and a graduated stem. All that is required with this instrument is, to pop it into the urine, and read off the number which happens to be on a level with the surface of the liquid. If it chance to be 15 (Fig. 3, D), then the urine has a specific gravity of 1015. If, on the other hand, the instrument is high in the liquid, and the reading is at 30, the urine is proportionally heavy, the density being 1030.

Although this instrument does admirably for clinical pur-poses, it is not sufficiently exact for scientific research ; accordingly, when we desire to make scientific observations, we use a picknometer (Fig. 3, E), which, as shown in the wood-cut, is a small bottle, with a long stopper perforated by a capillary tube. When the stopper is accurately adjusted, the bottle holds exactly 20 c.c. of distilled water. To ascertain the specific gravity of the urine by its means, the bottle is filled to overflowing, accurately stoppered, wiped dry, and weighed ; and the difference between its weight when filled with distilled water and with urine indicates the specific gravity. For example, suppose the urine weighs 20·3 grammes, (c) while distilled water weighs only 20 grammes, the calculation would be to multiply the 20·3 by 1000, and divide by 20.

$$\frac{20 \cdot 3 \times 1000}{20} = 1015$$

Or simply multiply the decimal by 5, and the result is the last two figures of the specific gravity, 3 × 5 = 15.

The Temperature of the Urine.—Patients frequently say their urine scalds them, and this has led to the idea that the urine is hot ; whereas its temperature, even at the moment of being passed, is never higher than the temperature of the body. The urine, in reality, receives its heat from the body ; and, as the material warmed can in no case be hotter than the material which warms, the urine cannot possibly have a higher temperature than the frame. When patients complain of the urine scalding them, we may be assured of one of two things, either that the canal, or part of the canal, is in a state of inflammation, and has, consequently, its sensibility increased ; or, that the urine possesses more than its usual degree of acrimony. The normal temperature of freshly-passed urine is about 99° F. ; but in cases of acute rheumatism, pneumonia, scarlet fever, and such other diseases as raise abnormally the bodily temperature, the urine participates in the change, and its temperature may reach as high as 109°, in the coma of sunstroke, or even 112 5° F. in idiopathic tetanus. On the other hand, it may fall as low as 79°, in tubercular meningitis, or even 77° F. in mania just before death.

(c) Calculated in grammes or grains, the result is precisely the same.

QUANTITY OF URINE.

The amount of water expelled from the body varies exceedingly, not only as regards the relative quantity voided at each micturition, but also as regards the absolute amount passed in the twenty-four hours. From what has been previously said with regard to the different kinds of urine passed in the day, one can easily see the absolute necessity of collecting all the urine made by the patient during twenty-four hours, and analysing a sample taken from the whole. Formerly, the constituents of the urine were reckoned at so much per cent.; but subsequent research having shown that such a mode of analysis gave rise to most erroneous data, it has been thrown aside as worse than useless.

Indeed, when making a study of any particular disease, the mere calculating the quantity of urine passed in a single day is not sufficient, for, as will be immediately seen, there are many things which accidentally influence the quantity. In order to be exact, the urine should be collected and measured during at least two, if not four, consecutive days, and the average for the twenty-four hours calculated. Some writers (Vogel and Neubauer), think it advisable to measure the hourly excretion; but this can scarcely be done except in Hospitals, and fortunately for those in private practice, there is no necessity for such measurements.

If we merely wish to ascertain the presence or absence of any particular substance, such, for instance, as albumen or sugar, testing any of the urines passed during the twenty-four hours, is in general sufficient for the purpose. It is usually recommended in books to examine the morning's urine (urina sanguinis). This recommendation, however, is not unlikely to lead to error, and that, too, in the very cases where it is of the utmost importance to avoid it, namely, in those where the disease is in its early stage, and, consequently, where but little of the abnormal substance is present in the urine. In such cases, it is much better to employ the urina cibi (that passed from three to four hours after dinner is the best), as it invariably contains the greatest amount of either of these foreign substances. The reason of this I shall subsequently explain.

In all cases where we desire to make a quantitive analysis of any of the urinary constituents, healthy or morbid, we must not only have a sample of the twenty-four hours urine, but at the same time be aware of the absolute quantity that has been passed. For this purpose the collected urine may be either

FIG. 4.

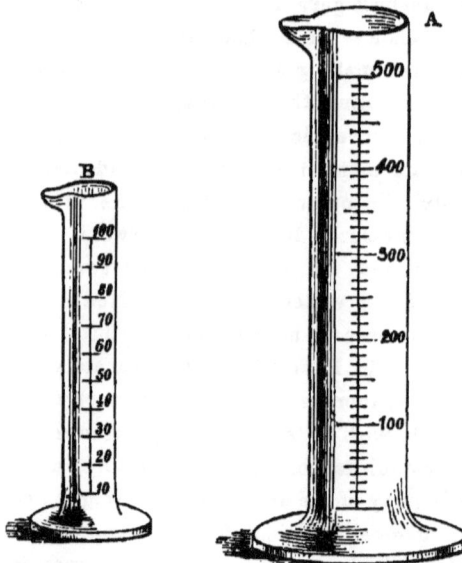

carefully measured in an ordinary ounce measure, and the number of cubic centimetres calculated therefrom, (b) or still

(b) The French measures are, when once learned, so very much easier to calculate than the English, that they are now being generally adopted. 1 ounce = 31 cubic centimetres; 1 c. c. or 1 gramme = 15½ grains of distilled water.

C

better, the urine may be measured in a glass jar, Fig. 4 (A),
or Fig. 4 (B), graduated in cubic centimetres, and the quantity thus at once exactly ascertained.

As there are certain physiological as well as pathological
conditions which influence the amount of water voided during
the twenty-four hours, it is necessary that I should at once
make you acquainted with the physiological conditions on
which the quantity and quality of the urine depends. These
are the drink, the kind of food, the state of the cutaneous and
pulmonary exhalations, the length of time the urine is retained
in the bladder, the condition of the stools, the sex, the age of
the individual, and the influence of remedies.

1st. *As regards the Influence of the Drink on the Amount of
Urine.*—The quanity of urine passed during twenty-four hours
is generally, though not always, in excess of the fluid ingesta.
Böcker found that when he drank 1260 c. c. (40½ oz.) his
urine amounted to 2621 c. c. (84½ oz.), and when he drank
3360 (108½ oz.) it reached the large amount of 4994 c. c.
(161 oz.) in the twenty-four hours. In each case you observe
the quantity of water ejected is considerably above the
amount of drink taken. Professor Vogel made some interesting observations on this point. He daily weighed the food
and drink of a person during 189 days, and found, as a rule,
that the kidneys gave off from $\frac{1}{20}$ to $\frac{1}{10}$ more water than the
fluid drunk ; (c) but it occasionally happened that only ⅓ of
the liquid taken was thrown off by the kidneys. The more
active the skin, *cæteris paribus*, the less water is excreted by
the kidneys. Whenever the quantity of drink is diminished,
the amount of urine passed is correspondingly decreased. We
next come to the important question of

2nd. *The Influence of Food upon the Amount of the Urine.*

(c) The largest quantity was passed after a cold bath, in consequence of
the combined effects of absorption of water and suppressed transpiration.

—This is not so easily understood as the influence of drink; it is, however, quite as remarkable. Various experiments have been made on this point, but I think the most interesting are those of Lehmann. They were made during the summer months, when the amount of water is much less than in winter, and the effect of different kinds of food was as follows:— While taking 1233 grms. (39·79 oz.) of animal food (eggs), the daily amount of urine excreted was 1202·5 c. c. (38·79 oz.) When living on vegetable food, which was certainly taken in larger quantity than the animal food, he passed only 909 c. c. (29·3 oz.) of urine. Lastly, when a mixed diet was taken, the amount of the urine stood between the two cited examples. This influence of food over the daily quantity of urine seems to depend chiefly on one of its elements, namely, the nitrogen. (d)

For example, Bidder and Schmidt found that a cat fed on animal food passed 45½ c. c. (1·46 oz.) of urine for every pound of its weight. Valentin ascertained that a horse fed on corn and hay passed only 6 c. c. (0·19 oz.) of urine for every pound of its weight (e) ; and I ascertained that a man on a mixed diet passed, on an average, 14 c. c. (0·45 oz.) for every pound of his weight; so here we see that, even in different species of animals, the rule regarding the influence of the kind of food over the daily quantity of urine, still holds good. Another equally interesting fact is, that it is not only the quality of the food, but also the quantity, that exerts an influence over the amount of water excreted by the kidneys. This I can scarcely better exemplify than by quoting the following experiment of Bidder and Schmidt :—

(d) 100 parts of animal food contain on an average 15 per cent. of nitrogen ; whereas the same amount of vegetable food contains only from 2 to 5 per cent.

(e) *Vide* " Budge's Physiology," sixth edition, p. 214.

A cat when fed upon

Grammes.	grains.			Per lb. of bodily weight.	
				c. c.	oz.
108	= 1674 flesh, passed	91	=	2·9 of urine.	
75	= 1162 ,,	,,	71	= 2·2	,,
44	= 682 ,,	,,	53	= 1·7	,,

Here the animal had drink *ad libitum;* and when the water was removed, though still receiving 44 grammes (682 grains) of flesh, the urine diminished to 26 c. c. (0·83 oz.) in the twenty-four hours.

Many animals on a flesh diet do not drink any water at all, and yet continue to pass fluid in far greater quantity than they can possibly receive with the food. In the winter of 1859, I kept six dogs during three months solely on animal food (boiled tripe), during which time, notwithstanding that they received no water, the average quantity of urine daily passed by each was 186 c. c. (6 oz.) They never appeared to be thirsty, and when offered water refused to drink it. Here we have another illustration of the fact, that more liquid may be excreted by the kidneys than is taken into the stomach. From whence, then, does this liquid come? The dogs just spoken of were kept in a cool atmosphere, and there can be no doubt that the fluid was absorbed by the lungs. In a hot and dry atmosphere, the lungs would have exhaled instead of inhaled moisture, and the result would have been, that the animals would have required water.

3rd. *Influence of the Cutaneous and Pulmonary Exhalations.* —The more rapid the pulmonary and cutaneous exhalations, the less the amount of urine excreted. Hence, we invariably observe that, after much exercise, the urine is scanty, dark coloured, and of high specific gravity. It is sometimes, indeed, so concentrated, after profuse perspiration, that it irritates the urethra. This counterbalancing influence of skin, lungs, and kidneys is of the utmost value in the treatment of renal disease. In acute nephritis, for example, when the suppression of urine is the result of renal congestion, and the patient is too weak to stand active treatment, the simplest and safest way to relieve the kidneys is to induce free cutaneous perspiration. A hot air bath in such cases will often save the life of the patient. The perspiration, be it borne in mind, does not only carry off water, but many of the urinary solids, inorganic as well as organic. Normal sweat contains uric acid, urea, phosphates, and chlorides; and in disease, as I shall afterwards show you, even the insoluble oxalate of lime may be excreted by the perspiration in such quantity as to cover the skin with a white crystalline crust. It is, therefore, easy to understand how the cutaneous exhalation may replace for a time the renal function.

That the influence of the pulmonary exhalation is nearly of equal importance, is well illustrated by the effect of a change of weather on the urinary excretion. All of you must have noticed that you pass much more urine in winter than in summer. In some cases, as much as double the amount of urine is voided in the cold as is passed in the hot season of the year, and this, to a certain extent, is independent of the cutaneous influence. The same remark is applicable to the kind of day: in a cold and damp day much more water is passed than in a warm and dry one, when the pulmonary exhalation is in full activity.

According to Wiederhold, (f) the lungs, like the kidneys and skin, excrete both organic and inorganic substances; for, on collecting a large quantity of expired air, and testing it with suitable reagents, he found that it contained chloride of sodium, uric acid, urate of soda, urea, and urate of ammonia.

4th. *The quantity of Urine passed also depends, to a great extent, upon the time it is retained in the bladder;* for, during its sojourn there, its aqueous particles, and, according to

---

(f) *Deutsche Klinik*, No. 18.

Kaupp (g), even many of its solids, are continually being . reabsorbed into the circulation. Any one can easily prove the first part of the proposition by making the following experiment upon himself. Let him resist a call to make water, and immediately proceed to take active exercise. In a very short time the inclination to micturate will pass away, and perhaps not return for an hour or two, when it will then be found that, instead of much, very little urine is voided, and that little is of a dark colour, and high specific gravity. In such a case, the water that was previously in the bladder must have been reabsorbed into the blood, and exhaled partly by the skin and partly by the lungs, and with it, no doubt, some of the volatile and more soluble of the urinary salts.

Since physiology has shown that the quantity of fluid present in the blood during health varies but slightly, and that this is in consequence of the renal secretion fluctuating in direct proportion to the increase or diminution of the pulmonary and cutaneous exhalations, the kidneys may, with perfect justice, be said to be the safety valves of the system, as they are assuredly the most important regulators of the amount of fluid present in the body.

5th. *Influence of the Stools.*—All that can be said on this point may be summed up in a few words. The greater the quantity of liquid passed from the bowels, the less is ejected from the bladder; so that, in cases of diarrhœa and dysentery, patients void much less urine than in a state of health.

6th. *Influence of Sex and Age.*—As a rule, males pass more urine than females, the proportion being about,—

Sex.	Urine in Twenty-four Hours.			
	c. c.	oz.	c. c.	oz.
Men . . .	1000 =	32 to	2000 =	64
Women . .	800 =	26 ,,	1400 =	45

(g) Kaupp found that when the urine is retained in the bladder the water is taken up most readily, then the phosphates, the chlorides, the sulphates, and, lastly, the urea.—(Parkes, p. 108.)

The influence of age is equally remarkable. Children, for their size (as Scherer first observed), pass proportionally more water than adults. I estimated the quantity of urine passed during four days by a girl aged eighteen months, and her mother aged twenty-seven years, and the result was, that the child passed 13 c. c. (0·42 oz.) for every pound of her weight, and the mother only 8 c. c. (0·26 oz.) for every pound of hers.

As age advances, the absolute, as well as relative, proportion of urine again diminishes. The amount of water passed by the kidneys appears, therefore, to be proportional to the metamorphosis of the tissues.

7th. *The Influence of Disease and Remedies.*—Disease greatly affects the quantity of the urine : in some few it is increased, as, for example, in polydipsia and diabetes ; in others, it is diminished, as in cholera, fevers, and inflammatory affections. Remedies have also a very powerful effect on the amount of water excreted by the kidneys. We all know how diuretic some medicines are—spiritus ætheris nitrici, for example—while others have just the contrary effect. Mineral substances —such as iron and copper for instance—lessen the quantity of the urine, while cantharides and arsenic almost totally arrest its secretion.

In a carefully-observed case, I saw the citrate of quinine and iron in the space of twelve days reduce the quantity of urine from 2325 c. c. (75 oz.) to 1860 c. c. (60 oz.) in twenty-four hours ; the specific gravity (1017) remaining the same.

Again, I have noticed in the same individual the quantity of urine still further reduced by means of conia from 1860 c. c. (60 oz.) to 1240 c. c. (40 oz.) in the space of twenty-two days ; but in this case the specific gravity rose to 1027, as the quantity of water diminished.

In another case of confirmed polydipsia occurring in a man aged fifty-six, the one whose urine is now on the table, and which is still under observation, the urine has diminished

1240 c. c. (40 oz.) in the space of six days, during which time
the patient has been taking the ammonio-citrate of iron in conjunction with quassia. The exact quantities of urine passed
by this patient are,—

Date.	Quantity of Urine in Twenty-four Hours.		Specific gravity.
	c. c.	oz.	
November 16 . .	4960 =	160	1010
November 22 . .	3720 =	120	1017

LECTURE II.

HAVING seen how the quantity of liquid excreted by the
kidneys varies, not only relatively at each micturition, but
absolutely, during the whole twenty-four hours, we have
next to consider how the relative and absolute amount of
solids excreted also vary, and to call attention to the causes
which induce the variations.

The solids consist of all that remains behind when the
aqueous part of the urine is driven off by evaporation. They
appear in the form of a brown-coloured, strongly-scented,
bitter-tasted mass, which, on analysis, is found to be composed
of acids and bases, organic as well as inorganic. Normal
human urine contains the following :—

	Acids.	Bases.
Inorganic	Hydrochloric.	Ammonia.
	Sulphuric.	Soda.
	Nitric.	Potash.
	Carbonic.	Magnesia.
	Phosphoric.	Iron.
	Fluoric.	Lime.

		Uric.	Urea.
		Hippuric.	Urohæmatin.
		Oxalic.	Creatinin.
Organic	.	Lactic.	Resin.
•		Malic.	Fat.
		Phenylic.	Albumen.
		Kryptophanic.(a)	Mucus. (b)

It was previously said, that the kidneys act the part of safety-valves to the system, in regulating the amount of fluid in the circulation ; they might with equal justice be called the " scavengers " of the frame, for not only do they eliminate the normally *effete* elements of the tissues, but also all foreign and noxious materials that chance to have found their way into the circulation. Hence it is, that we occasionally find a variety of substances in the urine, many of which cannot have played any part in the processes of life, as they are expelled from the

FIG. 5.

circulation almost as rapidly as they are absorbed. Ferro-cyanide of potassium, for example, has been detected in the urine within two minutes after it was taken into the stomach ; and many mineral and vegetable poisons are to be found in the urine within a very brief period after their admission into the system.

As regards Sediments, it may be here stated, once for all, that no spontaneous deposit takes place in normal urine within twenty-four hours after being voided. The only thing which is at all visible to the naked eye is a small quantity of mucus, which appears after six or seven hours as a white hazy cloud, suspended (not deposited) n the clear transparent urine.—Fig. 5. For

---

(a) Thudichum, Jour. Chem. Soc., 1870, p. 116.
(b) Some authors add to these bases manganese and creatin, and, to the

the purpose of demonstration, the mucus cloud is here represented much more distinct than natural; it ought only to be visible, and nothing more.

Whenever, therefore, any kind of deposit spontaneously forms within twenty-four hours, the urine may be regarded as abnormal. Either the proportion of some one or other of its solids is too great, or the proportion of its water is too little.

The effect of a mere diminution of the aqueous particles of the urine is easily rendered apparent, by simply concentrating a little of a perfectly healthy specimen. If, as soon as it is reduced to about one half of its bulk, it be placed aside to cool, a number of its solids will be precipitated in a crystalline form, the characters of which can be readily recognised by the aid of a microscope. This fact we occasionally turn to practical account when searching for substances which, like leucin and tyrosin, are only crystallisable in concentrated solutions.

PHYSIOLOGICAL AND PATHOLOGICAL CONDITIONS INFLUENCING THE EXCRETION OF URINARY SOLIDS.

There are several physiological as well as pathological conditions which cause the absolute quantity of solids passed in the twenty-four hours to vary in different individuals, and even at different times, in the same individual. In consulting books on the urine, you will be struck with the apparent discrepancies of French, English, and German writers as regards the quantity of solids passed by healthy individuals. Thus, an adult male on a mixed diet is said to pass on an average the following amounts:

	Amount of Solids in Twenty-four Hours.	
	grammes.	grains.
Frenchman (Becquerel)	39·52 =	612·56
Englishman (myself)	53·00 =	821·50
German (Lehmann)	67·82 =	1051·21

acids, taurylic, damolic, and damaluric; but they are so rarely found in healthy urine, that they scarcely deserve being noticed.

Paradoxical though it appears, each of these statements is in itself correct, the variance being simply due to the national peculiarity of the individual ; not that the French, German, and English constitutions are essentially different, but on account of the modes of life, and especially the diet of the individuals, being different. For example, Frenchmen take much vegetable, and little animal food, and what animal food they take is not of the most nourishing kind ; first, because it is originally inferior ; and secondly, because, in order to make it palatable, it is submitted to a complicated process of cookery, which still further decreases its nutritive value. Englishmen, on the other hand, eat moderately of both vegetable and animal diet ; while our German *confrères*, on whom the experiments have mostly been made, contrary to their own opinions, eat largely of all foods.(c) In order to show you that this is no imaginary statement, and that diet does exert the influence here attributed to it, I shall quote to you the results of an experiment made by a German (Lehmann) on himself :—

Diet.	Solids in Twenty-four Hours' Urine.	
	grammes.	grains.
Animal	87·44 =	1355·32
Mixed	67·82 =	1051·21
Vegetable . .	59·23 =	918·06
Non-nitrogenised .	41·68 =	646·04

Here is at once seen the immense influence diet exerts over the amount of solids excreted by the kidneys. They are, it will be observed, increased by an animal, diminished by a vegetable, and still further reduced by a non-nitrogenised diet ; and this, you will remember, is a precisely similar result

---

(c) As a nation, the Germans are much worse fed than the English ; but the students and Professors, on whom the experiments on healthy urine have been made, are a striking exception to this rule, as any one who has sat at a German *table d'hôte* must have observed.

to that noticed regarding the effect of diet on the quantity of liquid voided in the twenty-four hours. Curiously enough, the same rule holds good regarding the effect of drink on the solids; for, just as the amount of solids taken into the stomach influences the quantity of water eliminated by the kidneys, so the amount of water taken into the stomach influences the amount of solids excreted by the kidneys.

For a long time, Lecanu denied that with an excess of drink there was an augmentation of solid matter passed with the urine; but the researches of Winter, Vogel, and others, have placed the fact beyond a doubt. To me it appears quite natural that it should be so; not because I imagine that the fluids drunk augment to any great extent the metamorphosis of the tissues, but because, by increasing the quantity of fluid in the intestines, more food is dissolved, and consequently more solid matter absorbed into the circulation, and the excess beyond what the wants of the system require is excreted by the kidneys.

This view is founded on the result of an experiment made by Kierulf, who found that, after injecting water directly into the circulation, the solid matters of the urine were very little increased, although the urine became albuminous. Whereas, had the water which was injected directly into the blood been administered by the stomach, the urinary solids would have been proportionally augmented. Of course, no one would seek to deny that an excess of water in the circulation accelerates, to some extent, the metamorphosis of the tissues; all that it is desired to prove is, that this is neither the only nor the chief cause of the augmentation of the urinary solids after the free use of liquids.

It must be borne in mind that the foregoing remarks have only reference to the amount of solids eliminated in the twenty-four hours; for the proportion of solid matter excreted at each micturition is, in reality, reduced after free drinking.

This is easily understood when we remember that there are three kinds of urine passed during the day,—the urina sanguinis, urina potûs, and urina cibi,—and that these differ in specific gravity ; the difference being due to the variable amount of solids they contain. For example, if there are 46·6 grammes of solids in every 1000 cubic centimetres of the twenty-four hours' urine, possessing a specific gravity of 1020, there will be in every

Cubic Centimetres.	Specific Gravity.	Grammes.	Grains.
1000 of urina cibi . .	1025	58·25 =	902·87
„ „ sanguinis .	1017	39·61 =	613·95
„ „ potûs. . .	1009	20·97 =	325·03

So that when we speak of the drink augmenting the urinary solids, we mean augmenting the total quantity excreted in the twenty-four hours.

INFLUENCE OF AGE AND SEX ON THE URINARY SOLIDS.

Age has evidently an important influence on the urinary solids ; and although as yet few observations have been made on this point, they, nevertheless, all tend to one conclusion, namely, that children excrete, proportionally, more solid matter than adults. For instance, I found that, while a female child, aged 18 months, excreted 0·58 gramme (8·99 grains) for every pound of her weight, the mother, aged 27 years, living under the same conditions, excreted little more than half that amount (0·31 gramme = 4·80 grains) for every pound of hers.

Mother, aged 27; weight, 140 lbs.

	First Analysis.		Second Analysis.		Third Analysis.		Total Average.		Solids per lb. of bodily weight.	
	grammes.	grains.	grammes.	grains.	grammes.	grains.	grammes.	grains.	grammes.	grains.
Total solids in twenty-four hours . .	45·64 = 707·42		43·94 = 681·07		43·66 = 676·73		44·41 = 688·35		0·31 = 4·80	
Organic substance . .	31·73 = 491·81		32·04 = 496·62		33·78 = 523·59		32·51 = 503·90		0·33 = 3·56	
Inorganic salts	13·91 = 215·60		11·89 = 184·29		9·87 = 111·98		11·89 = 184·?9		0·08 = 1·24	

Female Child, aged 1½ year; weight, 21 lbs.

	Average of First and Second Analysis, calculated on Forty-eight Hours' Urine.	Third Analysis.		Total Average.		Solids per lb. of bodily weight.	
	grammes. grains.	grammes.	grains.	grammes.	grains.	grammes.	grains.
Total solids in twenty-four hours . .	12·40 = 192·20	12·04 = 186·62		12·22 = 189·41		0·58 = 8·99	
Organic substance . .	7·42 = 115·01	6·59 = 102·14		7·00 = 108·60		0·33 = 5·11	
Inorganic salts	4·98 = 77·19	5·44 = 84·32		5·21 = 90·75		0·24 = 3·72	

The influence of sex is no less remarkable than the influence of age, as is seen by the subjoined table, representing the average relative proportion of organic and inorganic salts daily eliminated by an adult English man and woman. The table is made from an average of four analyses :—

In Twenty-four Hours' Urine. Average weight, 140 lbs.; age, 25 years.

	Males. grms. grains.	Females. grms. grains.
Total solids	53·00 = 821·50	44·50 = 689·75
Organic	36·60 = 567·30	31·00 = 480·50
Inorganic	16·40 = 248·70	13·50 = 209·25
Solids per lb. of bodily weight	0·37 = 5·73	0·35 = 5·42

Women, it appears, therefore, excrete, both absolutely and relatively, less solid matter than men. (d)

INFLUENCE OF PREGNANCY.

Of all physiological conditions, that of pregnancy has the greatest effect upon the daily excretion of solid matter by the kidneys. The following observations were made on a primipara, aged 26 years :—

Period of Gestation.	Total Solids.	Organic.	Inorganic.
	grms. grains.	grms. grains.	grms. grains.
Two months and a-half	57·99 = 898·84	42·88 = 664·64	15·11 = 234·20
Fifth month, ten days after quickening ..	57·61 = 892·95	40·55 = 628·52	17·06 = 264·43
Seventh month ..	45·64 = 707·42	31·73 = 491·81	13·90 = 215·45
Ninth month, three days before delivery	43·94 = 681·07	32·04 = 496·62	11·89 = 184·29

Here it is clearly seen, that the nearer we approach delivery the less solids are excreted, and there can be little doubt but that this arises from the fact, that the more developed the

---

(d) Parkes gives the mean solids at 61·14 grammes, 945 grains ; but this arises from his including the high average of the Germans.

fœtus the more nutritive material it withdraws from its mother's circulation, and consequently the less remains to be excreted by the kidneys. It is to be remembered, too, that the effete products of the fœtal tissues are not directly returned to the maternal blood to be excreted along with her own, but are eliminated by the fœtal kidneys in the shape of uriñe, which is expelled from its bladder into the liquor amnii. This is, in fact, the true explanation why urea is so often detected in the amniotic fluid.

Influence of Disease on the Daily Excretion of Urinary Solids.

Disease exerts a marked influence over the urinary solids, its most general effect being rather to diminish than to augment them. In cases of anæmia and chlorosis, for example, the diminution is sometimes very considerable. The twenty-four hours' urine of M. L., aged 30 (Hospital case), suffering from anæmia, consequent on hyper-lactation, contained—

Total Solids. grammes. grains.	Organic. grammes. grains.	Inorganic. grammes. grains.
34·89 = 540·79	26·63 = 412·76	8·26 = 128·03

Lest some may suppose that in this case the diminution in the amount of urinary solids may have been due to a diminished supply of food, I shall quote another case from private practice, about which there can be no doubt. This patient was also suffering from anæmia, and being of the same size, and nearly the same age (32) as the preceding, it may be taken as a parallel case. Her diet was of the most nourishing description. Turtle, and other nutritive soups, well-cooked fresh meats, and three or four glasses of wine daily. Notwithstanding this, the twenty-four hours' urine yielded on analysis only—

D

Total Solids.		Organic.		Inorganic.	
grammes.	grains.	grammes.	grains.	grammes.	grains.
32·64 =	505·92	24·98 =	387·19	7·6 =	117·80

being even less than the Hospital case.

I have now to quote an example of an increase of the urinary solids as the effect of disease. The most interesting case of this kind which has fallen under my personal observation is that of a gentleman aged 43, of very high literary attainments. He had enjoyed perfect health till within eighteen months before I saw him, when, after great and continued mental exertion, he was seized with intense frontal headache, and violent palpitation of the heart; so much so, that he had to resign his appointment, and desist from all mental labour. When brought to me by Dr. King, he was suffering from other symptoms in addition to the above, the most distressing of which were, pain from the hips downwards, with shooting, burning sensations in the soles and upper part of the feet. He had also pricking, burning sensations in the wrists and hands, and to such an extent did these annoy him, that they prevented his ever enjoying a comfortable night's rest. On analysis, his twenty-four hours' urine was found to contain a very great excess of solid matter ; and this, too, in spite of his taking very little food, in consequence of his suffering from dyspepsia. Under treatment, a marked diminution of the urinary solids took place, and with it a corresponding improvement in health. The twenty-four hours' urine was analysed on three separate occasions, with the following results :—

	Total Solids.		Organic.		Inorganic.	
	grammes.	grains.	grammes.	grains.	grammes.	grains
March 19	80·67 =	1250·38	57·42 =	890·01	23·25 =	360·37
„ 31	71·61 =	1109·95	49·22 =	762·91	21·39 =	331·54
April 22	63·24 =	980·22	47·82 =	741·21	15·42 =	239·01

· At the time of the last analysis the patient was very much improved.

The quantitiy of solids daily excreted may, in certain diseases, be far greater than even the above.

In cases of diabetes, for instance, the amount of solid matter may be enormously increased—not merely on account of the presence of sugar, but also on account of several of the urinary ingredients being augmented. Here is the analysis of the urine of a gentleman, aged 57, who suffered for six years from diabetes. I quote it, not because it is the most striking example I could give, but because it happens to be the first that has come to hand, and because I shall have afterwards to refer to the case when on the subject of treatment :—

	Total Solids.		Organic.		Inorganic.	
	grms.	grains.	grms.	grains.	grms.	grains.
24 hours' urine . .	276·21	= 4281·25	256·60	= 3977·30	19·60	= 303·80

I might go on quoting many more such examples ; but these are sufficient for the present purpose. I shall, therefore, turn to the important question,—What effect have remedies in controlling the elimination of solids by the kidneys?

EFFECT OF REMEDIES ON THE URINARY SOLIDS.

There are certain remedies which have a powerful effect in reducing, and certain others which have an equal effect in augmenting the urinary solids. First, as regards those remedies which reduce them. Nearly all belong to the narcotic group ; and the most active are opium, morphia, conia, hyoscyamus, and Indian hemp. As one and all of these, from their specific action on the nervous system, are occasionally counter-indicated, it fortunately happens that we have another group of remedies of an entirely opposite neurotic character, which also have the power of diminishing the elimination of solid matter by the kidneys. The most powerful of these, I believe, is the citrate of quinine and iron, and next to it the ammonio-citrate of iron in combination with quassia —two remedies which, as was before seen, have also the property of diminishing, in certain cases, the amount of liquid excreted by the kidneys.

Those medicines which, on the other hand, increase the solids are digitalis, atropine, and colchicum—indeed, it is possible that the chief benefit derived from the use of these remedies in certain diseases arises from this action. In gout, for example, both immediately before and during the paroxysm, the urinary solids are said to be diminished; whereas, as soon as the paroxysm begins to subside, they are said to increase, and the more rapid the increase the quicker appears to be the convalescence of the patient. Now, it so happens that both the carbonate of potash and colchicum, which, as is well known, are useful remedies in the majority of cases of gout, increase the elimination of solid matter by the kidneys.

There are yet other substances which, although not usually classed under the head of medicines, nevertheless merit our attention. These are alcohol, wine, and beer—all of which influence the daily amount of solids passed by the kidneys. Thus, Böcker found that alcohol and beer lessen the quantity, while white Rhine wine (Niersteiner) augments it. Here, then, we see that in the treatment of disease we can even select our stimulants so as to further the effects of our other remedies.

INORGANIC SALTS.

It must have been observed that I have hitherto spoken of the urinary solids as a whole, notwithstanding that they consist, as you are aware, of two parts, an organic and an inorganic. I have purposely done so in order to save time; for, as a rule, what influences the one influences the other.

There are, however, certain exceptions to it, with which I must now make you acquainted.

From the preceding tables you will have learned that the organic is usually greatly in excess of the inorganic matter; and, although this is the case at all periods of life, yet the relative proportion varies with the age. Thus, it will be seen by the subjoined table, that while a child aged eighteen months passes for every pound of its weight only one-third more organic matter than the adult, it excretes no less than three times more inorganic salts :—

Solids per lb. of Bodily Weight in Twenty-four Hours' Urine.

	Total Solids.		Organic.		Inorganic.	
	grammes.	grains.	grammes.	grains.	grammes.	grains.
Girl aged 18 mths	0·58 =	8·99	0·33 =	5·11	0·24 =	3·72
Mother aged 27 yrs	0·31 =	4·80	0·23 =	3·56	0·08 =	1·24

The excess is here still seen to be in favour of the organic ; and even in disease this rule holds good. There is, however, one remarkable exception to it, which, as far as I am aware, has not been noticed by other observers. And this is the more surprising, seeing that not only is the relationship altered, but actually reversed, the inorganic preponderating over the organic ingredients of the urine. The most striking illustration that I can give of this most important fact is in albuminuria, and I think that I cannot do better than quote the one cited in a memoir " On the Blood and Urine of Pregnancy," published conjointly by Gegenbaur and myself in Scanzoni's "Beiträge" for 1854. The case was that of a woman, aged thirty-nine, the analysis of whose urine yielded the following results :—

In the 24 hours' Urine.	26 days before Delivery.		14 days before Delivery.		7 days after Delivery.	
	c. c.	oz.	c. c.	oz.	c. c.	oz.
Water . .	1169 =	37·7	1668 =	53·8	1792 =	57·8

grms.   grains.   grms.   grains.   grms.   grains.
Organic solids . 16·37═253·73   21·67═335·88   27·69═429·19
Inorganic salts . 28·88═447·64   26·70═413·85   31·01═480·65

Here it is seen that the inorganic ingredients preponderated over the organic on the twenty-sixth day before delivery as seven to four, on the fourteenth day as nine to seven, and on the seventh day after delivery, when the albuminuria was already beginning to diminish, as ten to nine.

DETERMINATION OF THE WATER AND SOLIDS.

Having ascertained the exact quantity of urine passed in the twenty-four hours, we can form a rough estimate of the amount of solids it contains by multiplying the last two figures of the specific gravity by 2·33. Thus, in the case of a urine of 1020 specific gravity, if the 20 were multiplied by 2·33, the result would be—the solids in 1000 cubic centimetres of urine 20 × 2·33 = 46·6 grammes.

If, instead, however, of 1000, 1250 cubic centimetres of urine were passed in the twenty-four hours, an additional "simple proportion" calculation would be required to ascertain the total amount. Thus, if 1000 cubic centimetres of urine contained 46·6 grammes of solids, 1250 cubic centimetres would contain 58·25; the calculation being to multiply 46·6 by 1250, and divide by 1000.

$$\frac{46·6 \times 1250}{1000} = 58·25 \text{ grammes.}$$

Again, if instead of 1250 cubic centimetres of urine being passed, there were only 800 cubic centimetres, the calculation would still be made on precisely the same principles.

$$\frac{46·6 \times 800}{1000} = 37·28$$

Although this method does very well when we merely wish to have a rough guess at the amount of solids passed in the twenty-four hours, and it might even be sufficiently exact for ordinary clinical purposes if the specific gravity of the urine

were carefully taken with a correct apparatus—the picknometer (Fig. 3, E), for example ; yet it is found to be in-
adequate for scientific purposes, and, therefore, it is necessary
to proceed as follows :—Twenty cubic centimetres of urine are
exactly measured off in a graduated pipette (Fig. 6, A), put
into a clean porcelain crucible of known weight, and again
weighed to ascertain the weight of the urine it contains,
should that not have been previously done in finding its exact
specific gravity. (I generally use the picknometer as the
measure, and thereby make one weighing do for both purposes.)
The uncovered capsule (Fig. 6, E) is now placed in a close

FIG. 6.

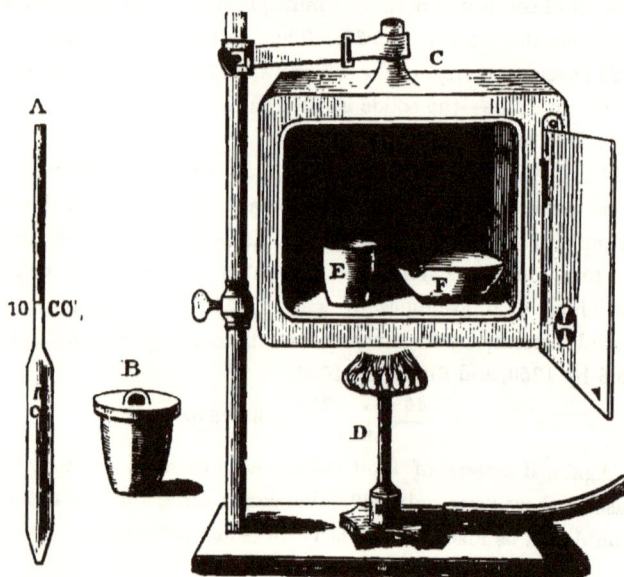

water-bath, heated to 100° C. (212° F.), and kept there until
the aqueous part of the urine is all driven off. When such is
the case, the crucible will cease to lose weight. (The drying is
further insured by placing the capsule for a short time under

an air-pump.)   The difference between the capsule when first
weighed and in its now dry state gives the quantity of water
in twenty cubic centimetres of urine.   If the patient chance
to have passed 1000 cubic centimetres in the twenty-four
hours, all that is required is to multiply the loss of water by
fifty, and the result is the total amount of water in the twenty-
four hours' urine.

Next, as regards the solids, the weight of the brown residue
now remaining in the crucible is the weight of the solids in
20 cubic centimetres of urine ; and supposing, as we have just
said, 1000 cubic centimetres of urine were passed, and the
residue weighed 0·6 grammes, the calculation would be :—
Multiply 1000 by 0·6, and divide by 20, the amount of urine
employed in the analysis—

$$\frac{1000 \times 0·6}{20} = 30$$

30 grammes, or 465 grains in the twenty-four hours. The calculation might be simplified by multiplying 0·6 by 50.

Lastly, to ascertain the amount of inorganic salts.   This is
done by placing the crucible containing the urinary solids over
the flame of a spirit-lamp until all the organic matter is
destroyed, and there remains nothing but the inorganic ash of
the 20 cubic centimetres of urine.(e)   The capsule is now
allowed to cool, and again weighed, when the weight of its
contents will represent the amount of inorganic salts in the 20
cubic centimetres of urine.   Suppose, for example, that they
weigh 0·3, the calculation would be—

$$\frac{1000 \times 0·3}{20} = 15$$

15 grammes, or 232·5 grains of inorganic substance in the
twenty-four hours' urine.

(e) In order to get the ash perfectly white, it is necessary to add either
a few drops of nitric acid, or a fragment of the nitrate of ammonia to the
charred mass, and again expose the crucible to a red heat.

LECTURE III.

UREA—ITS CHEMISTRY, PHYSIOLOGY, AND PATHOLOGY.
AMMONÆMIA AND URÆMIA.

We have now to consider the constituents of the urine individually, and we shall begin with that substance which, in consequence of its occurring in greatest quantity, is looked upon as the most important ingredient of the human urine.

Urea.—($C_2H_4N_2O_2$).   Atomic equivalent = 60.

Urea is to be found in the urine of all animals, but exists in greatest quantity in that of the mammalia.   It never occurs as a urinary deposit, in consequence of its great solubility in water, being, in fact, deliquescent.

Before entering upon the physiological, and pathological relations of urea, it will be necessary for me first to say a few words regarding its appearance and properties.

The urea of commerce is a white, semi-transparent, crystalline body, (a) with a bitter taste.   When heated it melts,

---

(a) The crystals are usually four-sided prisms, more rarely needle-shaped. The urea sold in the shops is prepared artificially, but it possesses the same characters as that obtained from human urine.

burns, and gives off ammonia. It is soluble in alcohol, as well as in water, but nearly insoluble in ether. It is readily decomposed by strong mineral acids, and by hydrated alkalies. With chlorine it yields carbonic, and hydrochloric acids. Nitrogen gas being at the same time set free. In contact with animal ferments urea takes up two equivalents of water, and is converted into two parts of carbonate of ammonia.

$$C_2H_4N_2O_2 + 2HO = 2(CO_2NH_3).$$

Hence the peculiar ammoniacal odour of putrid urine.

Urea readily combines with nitric acid to form the most important of its salts :—

The Nitrate of Urea (UrNO$_5$HO), which forms, when rapidly crystallised, flat, shining, rhomboidal plates; when slowly crystallised, fine prisms. It is very soluble in both water and alcohol, but only sparingly soluble when they contain an excess of free nitric acid.

We have here three specimens of urine—man's, horse's, and dog's—representing the three subdivisions of the animal kingdom into omnivora, herbivora, and carnivora; the first two urines have been concentrated to about half their normal bulk, the third is left in its natural state, it being by far the richest in urea. To each I now add equal parts of strong nitric acid, when instantly you perceive they begin to crystallise, and before many minutes elapse each of the liquids will have become one compact mass of shining crystals ; I shall be able to invert the glasses without danger of spilling their contents. You observe the crystals are coloured. Those from the dog's urine only slightly, those from the horse's very deeply. This is not owing to the nitrate of urea being a coloured salt, but simply because it happens to be deposited in a coloured liquid part of the colouring matter of which has united with it; just as sugar—a white substance—when crystallised in a solution of carmine, yields red crystals. The higher coloured the urine, therefore, the deeper coloured the crystals; the paler the urine the whiter the crystals.

With oxalic acid urea forms an oxalate (UrC_2O_3,HO) which crystallises in prisms and quadrilateral tables (Fig. 7), sometimes separate, sometimes adhering together in groups. This

FIG. 7.

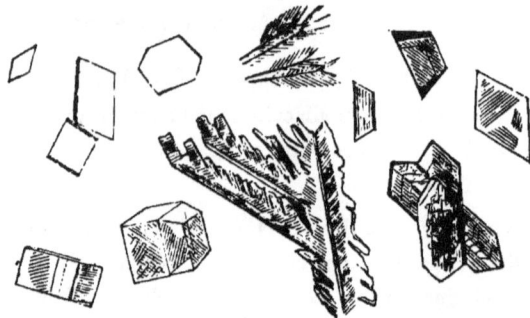

FIG. 7.—Crystals of oxalate of urea from human urine.

salt is freely soluble in hot, but sparingly soluble in cold, water. Hence it is that we occasionally, though very rarely, detect crystals of oxalate of urea in concentrated human urine.

PHYSIOLOGICAL RELATIONS OF UREA.

Urea was first discovered in human urine, and it was for a long time thought to be one of the special products of the animal world. Modern chemistry having, however, succeeded in preparing it artificially from a variety of substances, among which may be mentioned the cyanates of lead and silver; together with the fact of its being also formed by the union of cyanic acid and ammonia—

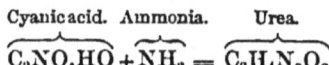

Cyanic acid. Ammonia. Urea.

$$C_2NO,HO + NH_3 = C_2H_4N_2O_2$$

many persons are inclined to regard it as merely the cyanate of ammonia.

Urea, though found in the urine, is not formed in the kidneys, but only excreted by them from the blood, where it has been detected by Garrod, myself, and others. Picard found 0·016 per cent. in normal human blood, and on one occasion I found as much as 0·320 per cent. In the defibrinated blood of a sheep I found 0·559, and in that of the dog as much as 0·747 per cent. In a normal state the blood of the renal artery contains twice as much urea as that of the renal vein (Picard), but in cases of its retention, both vessels contain about an equal quantity. Wurtz obtained urea from the healthy chyle, and lymph of the ox, horse, and dog. It has been found in the humours of the human eye, and is one of the ordinary constituents of the amniotic fluid of all mammalia; but, as I said before, it owes its presence in the latter situation to the urine excreted by the fœtus while still in utero.

It has been asserted by some (Robin and Verdeil) that no urea is to be found in the urine of very young children ; this, however, is an error, as I have detected urea in urine taken from the bladder of a still-born child, and in that of an infant aged 8 weeks I found as much as five grammes per thousand. In fact, from the subjoined table it will be seen that children actually excrete relatively more urea than adults ; and males absolutely as well as relatively more than females, at all periods of life.

Quantity of Urea in the 24 hours' Urine.

Sex and age.	Absolute amount.		Relative proportion per lb. of bodily wght.	
	grammes.	grains.	grmes.	grains.
⎧Boy aged 18 mths.	8 to 12 =	124·0 to 186·0	0·40	6·2
⎩Girl ,, ,,	6 to 9 =	93·0 to 139·5	0·35	5·4
⎧Man aged 27	. 25 to 35 =	387·5 to 542·5	0·25	3·8
⎩Woman ,,	. 20 to 30 =	310·0 to 465·0	0·20	3·1

In this table it will be seen that the amount of urea daily excreted is liable to considerable variations. Before attempting, therefore, to draw any conclusions from the changes observed in its amount during disease we must make ourselves thoroughly acquainted with the causes upon which the normal fluctuations depend, and in order to do this the more effectually we must first inquire into its normal source.

No urea has as yet been detected in any of the human solids, except what might readily be accounted for by the presence in them of blood. Its source seems to be two-fold; firstly, the disintegration of the tissues; and, secondly, the excess of nitrogenised food absorbed into the circulation.

The effect of food on the daily elimination of urea is well marked in the results obtained by Von Franque while experimenting on himself:—

UREA.

Diet.	In 24 hours.		Per hour.		Per lb. of bodily weight.	
	grmes.	grains.	grmes.	grains.	grmes.	grains.
Animal (3½ lbs. of flesh).	92 =	1416·0	3·86 =	59·8	0·53 =	8·2
Mixed	37 =	573·5	1·58 =	24·4	0·21 =	3·2
Vegetable	28 =	434·0	1·08 =	16·7	0·15 =	2·3
Non-nitrogenous	16 =	248·0	0·69 =	10·6	0·09 =	1·3

It is here seen that on an animal a much greater quantity of urea is excreted than on a mixed diet, and least of all after non-nitrogenised food has been taken. These results are in accordance with those of all other observers, except that the total quantities are greater.

Some suppose that urea comes in no case directly from the excess of absorbed food, but only indirectly from that portion of it which has undergone metamorphosis in the tissues (Bischoff, Voit, Parkes). If such be in reality the case, it appears to me very extraordinary that when living on a purely animal diet a person should excrete a larger proportion of urea than when living on mixed food, the mixed being the most nourishing diet of the two. Again, if their views be correct, why is it that exercise, which is well known greatly to increase the metamorphosis of the tissues, is not found mate-

rially to affect the daily elimination of urea? Thus, Von
Franque excreted in twenty-four hours :—

Diet and exercise.	Urea in 24 hours.		For every lb. of wght.	
	grammes.	grains.	grammes.	grains.
Mixed, with moderate exercise . . .	37·983	= 588·73	0·218	= 3·379
Mixed, with much exercise	37·877	= 587·09	0·217	= 3·363

Actually less urea after much exercise than he did on his
ordinary average exercise. (b)   On the other hand, Dr. Ed.
Smith found that prisoners excrete 36 grains more of urea
when at hard work than when doing nothing, which, he adds,
is 19 grains above what they pass on light labour.

Again, we find that Bischoff in his experiments on dogs
observed that they passed most urea when fed on eggs, muscle
and gelatinous substances, and that it increased in proportion,
to the quantity of food given.   These facts all tend to the
conclusion that a portion of the urea excreted by the kidneys
comes from the excess of food, independent of the metamorphosis of tissue.   That all the eliminated urea does not come
from this source no one seeks to affirm, for there is abundant
evidence against such a supposition.   For example, starve an
animal as you may the urea will never entirely disappear from
the urine.   It can only be reduced to a certain point and no
further.   Schmidt, Frerichs, Bischoff, and myself have
observed this in dogs, and Lassaigne, Scherer, Von Franque,

(b) So contradictory are the results of different observers, that further
experiments are necessary before the question can be settled ; the observations, too, in order to be exact, must only be made on the urine during
the time the exercise is being taken, and not over the whole twenty-four
hours, as the effect of rest and sleep may counterbalance the result
produced by the exercise.   I am myself strongly inclined to the opinion
that exercise increases the elimination of urea, but cannot venture posi-
tively to assert it in consequence of the contradictory evidence furnished
us by so many observers.   (Draper, Lehmann, Franque, Smith, Hammond, Ranke, etc.)

and others have made similar observations on the human urine. The last-named gentleman starved himself during forty hours, and still found 19·35 grammes (299·9 grains) of uréa in his twenty-four hours' urine. On experimenting on the effects of tartrate of antimony as a slow poison, I found·that the urine of dogs, reduced to perfect skeletons, contained such an amount of urea that it crystallised on the simple addition of nitric acid without any concentration whatever.

The amount of urea passed by an animal during starvation is greater, if he has been previously well fed, than if he has only received a sufficiency of food to prevent his losing weight (Bischóff), which fact probably accounts for Dr. Von Franque passing so much during the time he was without food.

Prout thought that the urea came chiefly from the gelatinous tissues, and Von Franque thinks it comes chiefly from the muscles, but neither of these tissues has been found to contain urea. It is true that Städeler and Frerichs found a considerable quantity of urea in the muscles of cartilaginous fishes; but that proves little, seeing that they detected it in almost all the organs. My idea is, that *urea is not the special product of any one particular tissue or organ, but the united product of all nitrogenised effete matter*. Bechamp and Picard assert that they have converted albumen and other azotised substances into urea by slow combustion with permanganate of potash, and although subsequent experimenters have not been equally successful, we must accept their statement as a fact until the contrary is proven.

Part of the urea may also be formed in the blood by the decomposition of uric acid. This is rendered highly probable, for two reasons—Firstly, Neubauer found on giving rabbits from two to three grammes (31 to 46½ grains) of uric acid with their food the quantity of urea excreted in the twenty-four hours was augmented from 2·1 to 4·2 grammes (32·5 to 65·1 grains); and, secondly, when uric acid is acted upon by permanganate of potash it is transformed into several substances, one of which is urea.

We have now to consider how the elimination of urea is influenced by other than animal foods. In the first place, water increases it. The more water we drink, *cæteris paribus*, the more urea we excrete, most probably for the reasons already given when speaking of the influence of water on the urinary solids. Common salt also produces an augmentation in the excretion of urea (Böcker, Bischoff). Coffee has an analogous effect. I found this to be very decidedly the case in one of my friends, Baron G., aged 57. The daily average of urea excreted by him was 17·28 grammes (267·84 grains), whereas after he had restricted himself to nothing but coffee (without milk or sugar) and dry bread during forty hours he passed 25·3 grammes (382·25 grains) of urea per twenty-four hours.

This diet had an equally striking effect on the amount of water eliminated. His usual average was 1290 cubic centimetres (41 ozs.), and on the day in question it rose to 1895 cubic centimetres (60 ozs.) per twenty-four hours. This result is contrary to the observations of Böcker and Hammond, who found, on taking an excess of coffee with their ordinary diet, that the daily excretion of urea diminished. But this probably arose from their taking the coffee in the usual way, sweetened with sugar, for there are certain foods which greatly diminish the elimination of urea, and among them sugar holds the foremost rank (Böcker, Bischoff, and Voit). Starch has also a marked effect in diminishing the amount of urea, and I believe that fat may be placed in the same category, for, notwithstanding the statements made to the contrary (Böcker and Bischoff), the results of Botkin's experiments on dogs are conclusive regarding the decided diminution of urea which takes place when fatty matters are added to the food. This opinion is further strengthened by the observations of Beneke, who found that cod-liver oil lessens the amount of urea passed by the human subject. In fact, my idea is, that all nitrogenised foods augment, and all non-nitrogenised foods diminish, the amount of urea eliminated by the kidneys. Even gelatine, the nitrogenised substance which of all others is supposed to possess the least nourishing properties, has a decided effect in increasing the daily excretion of urea.

I have entered thus fully into the physiology of urea in consequence of being aware that unless the influence of diet, age, and sex be borne in mind while studying its fluctuations during disease, our labours will be unattended with any practical benefit; while, on the other hand, if these factors are duly considered, we cannot fail to reap important advantage from the knowledge thus acquired, for just as we have seen that its amount in normal urine represents, to a certain extent, the wear and tear of the frame, and is consequently an index to the health of the individual, so also in disease it furnishes us with precisely similar and even more important data ; for the amount of urea in the urine enables us not only to judge of the tissue consumption, and, through it, of the severity of the affection, but is at the same time a direct clue to treatment, showing us whether our remedies should be directed to the restraining or encouraging tissue metamorphosis.

PATHOLOGY.

Disease has an important influence on the amount of urea excreted by the kidneys. In some diseases it is abnormally increased ; in others it is as strikingly diminished. Thus in all febrile affections, with one exception, namely, yellow fever, it is increased. In typhoid, for example, Parkes found as much as 57 grammes (883·5 grains) passed in the twenty-four hours ; and Vogel, in one case, the enormous quantity of 78 grammes

(1209 grains), all, or nearly all, of which, seeing that the patient was on low diet, must have come from the tissues. In typhus, febricula, and remittent, the urea is likewise found to be increased, though not to such an extent as in typhoid. The same remark is equally applicable to the different forms of exanthemata, as well as to inflammatory affections, such, for example, as pneumonia and meningitis. In the case of chronic cerebritis, already alluded to under a different head, I found the urea as high as 57·42 grammes (890 grains), and as the patient recovered it fell to 46·5 grammes (720·7 grains), and still later to 37·1 grammes (574 grains) in the twenty-four hours.

In some cases of epilepsy Dr. Sieveking found the urine so loaded with urea that the simple addition of nitric acid to the non-concentrated liquid caused an immediate formation of crystals of the nitrate. In one of his cases, in which I estimated the quantity, I found it amount to no less than 50 grammes per thousand. Unfortunately, no exact analyses have as yet been made on the twenty-four hours' urine of epileptics, but from the observations of Hensen it seems that although the urea may be increased during the seizure, it is in general diminished when estimated over the twenty-four hours.

Sanger found the daily excretion of urea diminished in cases of paralytic idiocy.

In a case of pyæmia, Vogel found 80 grammes (1240 grains) of urea ; and Ringer has made the interesting observation that the quantity excreted stands in a certain relationship to the temperature of the body ; for both in the hectic of phthisis, and in the paroxysms of ague it increases with the rise, and diminishes with the fall of the animal temperature. Scarlet fever is, however, an exception to the rule.

In diabetes the amount of urea is greatly increased ; on one occasion I found 70 grammes (1085 grains) passed in the twenty-four hours, and that too in the case of a gentleman nearly 50 years of age.

The average amount yielded by the twenty-nine analyses of diabetic urine now before me is 47 grammes (728·5 grains) —still a large excess over the normal excretion. But this is, I believe, in a great measure due to the rich animal diet. In a case of accidental diabetes accompanying paralysis, following upon a fall on the back of the head, instead of an excess there was an actual diminution of the urea. The man, a strongly built and otherwise healthy labourer, passed only 19·99 grammes (308·4 grains) of urea a few days after the receipt of the injury, while he was passing 36·89 grains of sugar; whereas during the period of convalescence, three months afterwards, when the sugar had disappeared from his urine, he passed 24·7 grammes (382·8 grains). The quantity of the urine in the first instance was 952 c.c. (30 oz.), of a specific gravity of 1017 ; and in the second, 915 c.c. (29 oz.), of a specific gravity of 1023.

In diabetes insipidus (polydipsia) I have found the urea enormously increased. On one occasion a man, aged 56, passed 84·63 grammes (1311·76 grains) in the twenty-four hours. At the same time the urine amounted to 4030 c.c. (130 oz.).

There are a few diseases in which the excretion of urea appears to be diminished before the attack. Gout (Böcker) and asthma (Ringer) may be cited as examples.

In cholera, the quantity of urea in the urine, both absolutely and relatively, is very small. The total quantity of urea excreted in the twenty-four hours, in some cases, does not amount to more than 4 grammes (62 grains). As soon as the attack begins to pass off, the urea commences to increase, and during convalescence it may even be much greater than in health. 70 grammes = 1085 grains (Buhl).

The question which here suggests itself is, whether the

decrease of urea in the urine of cholera is simply due to diminished elimination or to an absolute decrease in its formation? I am inclined to believe that the chief cause is the latter—viz., lessened formation, for while I can find no reason for its non-elimination, relatively to the amount of urine, if it be actually present in the blood, I see a potent one for its non-formation in the low temperature so characteristic of cholera. (We previously saw that in hectic and ague the urea rises and falls with the increase and diminution of animal heat.) Diminished formation cannot, however, be said to be the cause of the small amount of urea excreted in Bright's disease and some other affections.

In the experiments made conjointly by Gegenbauer and myself on the blood and urine in albuminuria, we found the daily excretion of urea on one occasion as low as 3·3 grm. (51 grains) and this was clearly due, not to diminished formation, but to arrested elimination. The more albumen excreted, the less was the amount of urea ; the less albumen excreted, the greater was the amount of urea in the urine.

In the 24 hours.	November 5. Grammes. Grains.	November 17. Grammes. Grains.	December 7. Grammes. Grains.
Urea .	. 3·3 = 51·15	5·4 = 83·70	10·9 = 168·95
Albumen	. 9·5 = 147·25	2·9 = 44·95	1·6 = 24·80

In cases of Bright's disease the urea is not only diminished in the urine, but is in excess in the blood. It may even make its appearance in all the fluids of the body. In dropsical exudations, in the humours of the eye, in the serum of the ventricles of the brain, in the milk, as well as in the vomited matters. On one occasion I found the blood of a patient who died of uræmic convulsions from Bright's disease quite acid (twelve hours after death), and in the serum taken from some large bullæ on the chest and thighs urea was detected in considerable quantity. It may here be mentioned that in yellow fever the urea, in like manner, appears to accumulate in the blood, so that in those cases it is not the formation, but only the excretion which is arrested. Dr. Porcher says that in some of the urines in the Charleston yellow-fever epidemic of 1846 there was an entire absence of urea. But this statement must be taken with a certain degree of reserve, as his mode of analysis was not sufficiently delicate to admit of his detecting small quantities of urea.

Most of the patients died with symptoms of uræmic poisoning.

URÆMIA.

Uræmic poisoning is a subject of such grave importance to the Medical man that I must say a few words regarding it. There can be no doubt that urea is a powerful irritant poison. When injected into the veins of animals it rapidly induces fatal convulsions. It is not necessary, as some have imagined, that the urea should be decomposed in the blood in order to produce its toxic effects. On the contrary, I believe that urea is a far more dangerous poison than the substance into which it is decomposed, namely, the carbonate of ammonia. Both are, however, poisonous, and they frequently manifest their action together, for although we may, perhaps, have uræmic without ammoniacal poisoning, we cannot have the latter, as the result of disease, without its being in some measure associated with the former.

True uræmia depends on the arrested elimination of the poisonous material by the kidneys; true ammonæmia on the re-absorption into the circulation of the decomposed secreted product.

Uræmia may occur in the course of any disease in which suppression of the renal secretion takes place. It is therefore superfluous to specify the diseases in which it may occur by name.

Carbonate of ammonia poisoning occurs in its least complicated form in those cases where the urea, although secreted, is retained in the urinary passages until it is decomposed, and the products of its decomposition are re-absorbed into the blood. Just as was formerly said (p. 22) the normal ingre· dients are re-absorbed from the bladder, when from any cause the urine is retained for some length of time in that viscus. The conditions under which it most commonly occurs are therefore :—

1st. Sacculated kidneys.

2nd. Dilatation of the ureter and pelvis of the kidneys.

3rd. Renal abscess.

4th. Paralysis of the bladder.

5th. Enlarged prostate with retention.

6th. Stricture with retention ; or, indeed, retention from any cause whatever,—all that is required to induce ammonæmia being the retention of the urine in the body sufficiently long to allow of the decomposition of its urea. It is quite surprising, indeed, in how very short a time retained urine may become ammoniacal. (c)

In Mr. Marshall's case of exfoliation of the bladder, which has during the last year given rise to so much discussion, the urine contained such an abundance of carbonate of ammonia that it effervesced exactly like a seidlitz powder when I added to it a few drops of acid.

The symptoms which distinguish ammonæmia from simple uræmic poisoning have been recently carefully pointed out by Jacksch (1860).

In ammonæmia the urine is ammoniacal when passed. The breath and perspiration are ammoniacal. The mucous membrane of the mouth is dry and shining. The complexion is sallow, and there is increasing emaciation. No dropsical

---

(c) Urine becomes much more rapidly ammoniacal in than out of the body, in consequence of the animal warmth favouring its putrefaction.

symptoms are present.   Convulsions are rare, but intermittent
ague is frequent. Moreover, although in the acute form of the
disease both vomiting and diarrhœa may occur, in chronic
cases, which are much more common, these symptoms are
always absent. Death is in general preceded by coma, varying
from a few hours to several days in duration.

It is thus seen that the symptoms, as well as the pathology
of ammonæmia, present striking features of difference to those
of pure uræmia. The most characteristic of the latter being
the vomiting, and diarrhœa, the convulsions, and coma. The
treatment of the two cases is in several particulars equally
different.

In the first place, ammonæmia arising, as it generally does,
from directly remediable causes, is much more frequently under
our control than uræmia. For example, when the ammonæmia is due to the simple retention of urine in the bladder
repeated catheterisation is usually followed by a speedy cure.
In no case is uræmia within the reach of instrumental interference. Its most remediable form is when it arises from some
temporary incapacity of the kidneys to perform their office ;
as in suppression of the urine after scarlet fever, or in idiopathic nephritis. Here we can do good by relieving the congestion of the kidneys by cupping or leeching the loins ; and
when the patient is too weak for such active measures, by the
application of dry cupping-glasses, poultices, and the free use
of the air-bath. In most other particulars, the two diseases
are to be treated alike ; and when neither offers us a chance of
permanent cure we can in general alleviate the sufferings, and
prolong the life of the patient by reducing the formation, as
well as by hastening the elimination of the urea. This may
be accomplished by a proper selection of food and medicine.

The remedies which enable us to control the formation, and
hasten the elimination, of urea are the following :—Whenever
there is an excessive tissue waste, as indicated by an augmented

excretion of urea, without any very apparent cause, the citrate
of quinine and iron appears to be a most useful remedy.
Sigmund has observed that digitalis diminishes the elimination of urea ; and Kletzinsky, that benzoic acid has a similar
effect, but in a much greater degree. The last-named observer
states that the daily excretion of urea may actually be reduced
to 2·5 grm. (38·75 grains), by benzoic acid, in the short space
of twenty-four hours. The acetate and phosphate of soda, as
well as colchicum, act in a similar manner (Parkes). On the
other hand, we have several remedies which increase it.
Thus, Sigmund found that cubebs and cantharides have this
effect ; and Parkes has noted that atropine augments its
elimination. If we wish to diminish the urea by means of the
diet, all that is necessary is to administer arrowroot, sago,
tapioca, and other such starchy foods, well sweetened with
sugar ; and when a more nourishing diet than this is demanded,
cream, cod-liver oil, or any other fatty matters that may be
considered suitable to the particular case, will have an equally
good effect in reducing the amount of urea. On the other
hand, when it is deemed advisable to increase it, animal soups,
eggs, milk, jellies, and other nitrogenised matters, together
with a fair portion of common salt, will answer the purpose.
To these may be added coffee, but without sugar.

In concluding my remarks on urea, I have to remind
you how useful a knowledge of its daily excretion is in
prognosis. Thus, it will have been gleaned from what has
already been said regarding the amount of urea in the urine
during the course of disease, that if, in cases of fever, such as
typhoid and typhus, or of the exanthemata, such as small-
pox and measles, or of inflammatory affections, such as pneumonia and meningitis, a decrease in the daily elimination of
urea is observed, it may be regarded as a most favourable sign
for the prognosis ; . for no sooner does a change for the better
take place in these affections, than an immediate diminution

in the amount of the urea is observable. During convalescence, the quantity is frequently below the normal standard. Whereas, in cases tending to a fatal termination, even in spite of the true febrile symptoms having passed away, the daily excretion of urea still remains high. Hence this sign may be a valuable and truthful guide when all others fail. If, on the other hand, in those diseases in which the excretion of urea is known to be abnormally small, such, for instance, as paralysis, cholera, or the different forms of albuminuria, an increase in its amount during the course of the case is noticed, it is an equally favourable sign for the prognosis; while, on the contrary, any further reduction in the amount of the eliminated urea cannot be otherwise regarded than a most untoward event, as, even when the other symptoms have improved, it is an almost infallible index of approaching danger.

Method of Estimating the Amount of Urea.

All that now remains for me to do is to show you the quickest and best method of ascertaining the quantity of urea contained in any given specimen of urine. This I believe to be Liebig's method,—for the employment of which furnish yourselves with the two following solutions, and prepared paper :—

1st. Liebig's standard solution of the proto-nitrate of mercury.(d)

Dissolve in strong nitric acid by the aid of heat 77·2 grammes (1196·6 grains) of dry oxide of mercury, sufficiently pure to leave no residue when volatilised on a platinum spatula. Evaporate the solution to the consistence of a syrup, and then dilute it up to 1000 cubic centimetres with distilled water.

(d) The standard solutions are to be obtained from Schacht and Hilgenberg, 38, Houndsditch ; Bullock and Reynolds, Hanover-street, Hanover-square.

If any precipitate of basic salt should form, add a drop or two more of the acid till it is re-dissolved. One cubic centimetre of this solution precipitates 0·01 gramme (0·15 grain) of urea.

2nd. Baryta solution.

Mix one part of a cold saturated solution of nitrate of baryta with two parts of a cold saturated solution of caustic baryta, and then reduce the mixture to one-half its strength by adding an equal amount of distilled water.

3rd. Impregnate a sheet of ordinary white filter paper with a saturated solution of carbonate or bicarbonate of soda ; dry, and cut it into strips six inches long and one inch broad. These may be kept in readiness in a wide-mouthed, stoppered bottle.

The next point is the requisite apparatus, the whole of which is represented in Fig. 8.

By using the above apparatus and proceeding in the manner I am now about to show you, one can easily make a quantitative analysis of urea in ten or fifteen minutes. I need not waste your time by explaining in what the superiority of the mode of procedure consists, as those who are in the habit of making these analyses will at once observe its practical advantages; while to those to whom the subject is new, it signifies not whether it is precisely the same or totally different from the ordinary methods, so long as it is the quickest and best.

Fill glass A as I now do to overflowing with urine ; empty it into one of the conical (C) glasses ; again fill it to overflowing with the baryta solution, which will precipitate all the sulphates and phosphates, and add it to the urine. Stir the mixture well, and then leave it for a few minutes at rest, while you take a filter, fold it in four, moisten it with distilled water, and insert it into the holder (D) placed over the other conical glass. Having done this, pour the mixture of urine and baryta solution into it, and while the filtering is going on fill one of the Burettes *(B)* with Liebig's standard solution to the level of the first graduation. Place ready on the stand one of the soda papers.

FIG. 8.

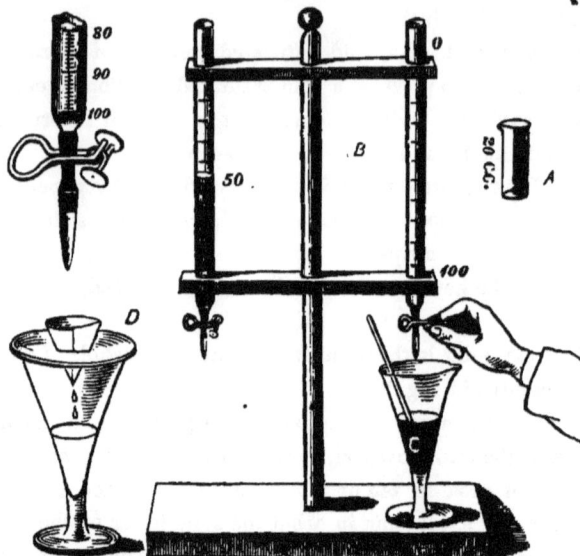

A. A glass measure holding exactly twenty c.c. when full to overflowing.
B. A stand containing two Burettes graduated in cubic or half cubic centimetres.
C.C. Two glasses, made specially of a conical form, in order to act as a support to
D. A glass filter-holder, like a candle ornament, only larger.
E. A filter.

You perceive that a considerable quantity of pure liquid has now filtered through the paper. You need not wait till the whole has passed through, but at once fill the *A* measure with the filtrate to overflowing, throw away the remainder, and return these 20 c.c. of mixture, which of course contain 10 of urine, to the glass. Now add to it a few drops of nitrate of silver, in order to precipitate the chlorides, for as long as any soluble chloride is present the

nitrate of mercury will give no precipitate with urea. (Many
do not remove the chlorides unless they are in large quantity,
as in ordinary cases they can scarcely be considered to interfere with the experiment.) I now add a drop or two of carbonate of soda solution to the liquid, in order to be perfectly
sure that it is alkaline—in acid solutions the precipitate of
mercury and urea (NO_5,Ur + 4HgO) does not form—and at
once proceed gradually to add to it the standard solution until
all the urea is precipitated, and there remains a slight excess of
the standard solution in the liquid. This is known by a drop
of the mixture producing a yellow stain on falling on the soda
paper. Whenever this coloration is detected, the analysis is
completed, and the quantity of urea in the twenty-four hours'
urine is ascertained by the following calculation :—You perceive that we have used 50 c.c. of the standard solution to
precipitate the urea in 10 c.c. of urine, and 50 c.c. of the
standard solution is equal to 0·5 gramme (7·5 grains) of urea ;
consequently as the patient passed 800 c.c. of urine in the
twenty-four hours, we multiply the 800 by 0·5, and divide the
result by 10, the quantity of urine analysed. The product
represents the quantity of urea excreted in the twenty-four
hours.

$$\frac{800 \times 0\cdot5}{10} = 40$$

namely, 40 grammes (620 grains).

N.B.—*Albuminuria.*—In albuminous urine the urea cannot
be estimated till after the albumen is removed. In order to
do so, acidulate the quantity of urine to be operated on with
a few drops of acetic acid. Boil it rapidly for a minute or
two over the spirit lamp, and then place it aside to cool.
When cold, the coagulated albumen will be found deposited
at the bottom of the vessel, and the clear supernatant liquid
can either be filtered or decanted. It is now ready to be
operated upon as above.

Albumen may also be separated from urine by passing the liquid through a filter filled with crystals of sulphate of soda. These have the power of arresting the albumen.

Diabetes.—The presence of sugar in the urine in no way interferes with the analysis of urea. •

Mucus.—When urine contains a large quantity of mucus or other deposit, only the clear supernatant liquid is to be employed in testing for urea.

LECTURE IV.

URIC ACID: ITS CHEMISTRY, PHYSIOLOGY, AND PATHOLOGY;
ITS PRESENCE AND SIGNIFICANCE IN DISEASES OF THE
HEART, LUNGS, LIVER, SPLEEN, AND KIDNEYS, IN GOUT,
AND SOME OTHER DISEASES, AND AS SAND, GRAVEL, AND
CALCULI.

Uric Acid $C_{10}H_4N_4O_6$; atomic equivalent, 168.

CHEMISTRY.

Uric acid is a white crystalline, feebly-acid, (a) tasteless,
organic substance; insoluble either in alcohol or in ether, and
only sparingly soluble in water. 2000 parts of hot, and
11,000 parts of cold water are required to dissolve one part of
uric acid. It is readily soluble in strong sulphuric acid, without undergoing decomposition, as may be proved by its speedy
re-precipitation on the simple addition of water.

By dry distillation uric acid is transformed into urea,
cyanuric acid, hydrocyanic acid, carbonate of ammonia,
aloxan, and an oily coal. Uric acid unites with bases, such
as soda, potash, and ammonia, and forms with them crystalline urates.

Both uric acid and its salts are easily recognised by yielding
the beautiful purpurate of ammonia (murexide of Liebig)
when treated with nitric acid, and ammonia. The best way
of applying the test is as follows :—Place a little of the

(a) It is so feebly acid that it is unable to redden litmus.

suspected substance on a platinum spatula or piece of white porcelain, moisten it slightly with strong nitric acid and heat the mixture gently over a spirit lamp, till the excess of acid is driven off. If uric acid or any of its salts be present the dry mass assumes, on the addition of a few drops of liquor ammoniæ, a brilliant crimson colour, and on the addition of caustic potash a purplish blue tint.

In employing this test the only substances likely to be mistaken for uric acid are tyrosine, hypoxanthine, and xanthoglobuline, which, according to Scherer, gave a similar reaction. With these substances, however, the colour is more of a yellow than a red, and after incineration there is left a black glistening stain on the platinum whereby they are distinguished from uric acid. Caffeine is said to give the murexide reaction, but it is a substance not likely to be confounded with uric acid.

Seeing that the murexide test is not always to be relied on in the presence of tyrosine, hypoxanthine, and xanthoglobuline, Schiff recommends freshly precipitated carbonate of silver to be employed. Nitrate of silver precipitated by carbonate of soda, one c.c. of which gives a grey colour with $\frac{1}{17600}$th gramme of uric acid. As chlorides and phosphates disturb the reaction, a few drops of nitrate of silver are to be added to the suspected liquid, quickly filtered, and then the carbonate of soda solution added ; the quantity of carbonate of silver precipitate ought not to be too great.

The reaction can be easily shown on filter paper in the following manner :—Dissolve a little uric acid in carbonate of soda or potash, put a drop or two on paper, and add nitrate of silver, a distinct grey stain instantly appears, thereby indicating the presence of uric acid. Tannic acid is the only other substance which gives the grey colour with carbonate of silver, and it can be distinguished from uric acid by the chloride of iron turning it black.

Quantitative Analysis.—To two hundred cubic centimetres
(6·45 oz.) of urine add twenty c.c. of strong hydrochloric or
nitric acid. (b)   Stir the mixture, and set it aside in a cold
place for twenty-four or thirty-six hours.   By the end of this
time the uric acid will have crystallised at the bottom, on the
sides, and at the surface of the liquid, which will have assumed
a dark colour in consequence of the liberation of the urohæmatine by the acid.   Collect the crystals on a small filter, wash
them slightly with alcohol, acidulated with a few drops of
hydrochloric or nitric acid, dry and weigh them.

The amount found is the quantity of uric acid contained in
200 c.c. of urine.   In order, therefore, to find how much has
been passed in the twenty-four hours, we say if 200 c.c. of
urine contain x grammes of uric acid, what will 1000 c.c. of
urine (or the amount passed in the twenty-four hours)
contain?

$$\frac{x \times 1000}{200} =$$ to the quantity of uric acid passed in 24 hours.

We have next to consider the physiological and pathological
relations of uric acid.

PHYSIOLOGY.

Uric acid is found in the liquid urines of the omnivora and
carnivora, as well as in the solid urines of serpents, birds, and
insects.   Almost the only urine in which uric acid does not
occur is that of the herbivora.   It exists in urine, combined
with soda, potash, and ammonia.   Its quantity varies greatly
in the urines of different species of animals.   For example,
while that of serpents is almost entirely composed of it, that
of the carnivora contains but a very small proportion.(c)

(b) Thudichum prefers the nitric, as uric acid is even less soluble in it
than in hydrochloric acid.

(c) I have frequently found it absent from dog's urine, when the
animals were entirely flesh fed.   In these cases the urea was always very
abundant.

Human urine stands between the two; but even in it the quantity of uric acid seldom exceeds one part per 1000 (Lehmann). Its relative proportion to that of urea is generally put down as 1 to 30.

During health uric acid is never absent from human urine, though the quantity varies very much according to the kind of food. In disease there is occasionally none to be detected.

Quantity of Uric Acid passed during Twenty-four Hours by Healthy Adults.

Diet.	Name of observers.	
	Lehmann.	The Author.
	grammes. grains.	grammes. grains.
Animal . .	1·478 = 22·9	1·250 = 19·3
Mixed . .	1·183 = 18·3	0·755 = 11·7
Vegetable . .	1·021 = 15·3	0·500 = 7·7
Non-nitrogenised	0·735 = 11·3	0·340 = 5·27

The above tables show that the quantity of uric acid daily excreted is greatly influenced by our food. It further appears that the amount depends on conditions analogous to those influencing the quantity of urea.

It will be observed that, although the relative proportion of uric acid passed by Lehmann and myself on different diets is tolerably uniform, yet the absolute amount is quite different. The same law seems to be applicable to uric acid as is applicable to urea, namely, that the daily excretion is much influenced by national peculiarity.

The following table illustrates this :—

On mixed diet.	grammes. grains.	Name of observers.
Germans . .	0·939 = 14·554	Kerner.
Englishmen .	0·755 = 11·702	The Author.(d)
Frenchmen .	0·510 = 7·905	Becquerel.

(d) Parkes puts the mean of uric acid passed by adults during the twenty-four hours at 0·555 grammes = 8·56 grains, and the proportion as 0·059 for every pound of bodily weight

F

While the influence of the kind of diet on uric acid and urea is much the same, that of the kind of drink is very different. Thus, port wine and beer increase the elimination of uric acid, whereas tea and coffee diminish it, which is the reverse of what happens with urea.

It is curious that the urine of human infants while at the breast is devoid of uric acid (Bird), while that of sucking calves possesses it (Wöhler). It seems, indeed, as if the natural order of things was reversed, for infants' urine is said to contain an excess of hippuric acid, and calf's urine none. Some may doubt the fact regarding the absence of uric acid from the urine of the human infant on account of the circumstance that uric acid calculi are not at all uncommon in very early life. This, however, arises from pathological conditions of the system, while the other is supposed to be due to physiological states. I have twice tested infants' urine—one aged ten days, the other eight weeks—for uric acid, and found none.

Climate has a most striking effect on the quantity of uric acid passed in the twenty-four hours. Nearly double the quantity is excreted in winter as in summer. Dr. G. Pringle, while working in my private laboratory, found that he passed, during summer, 0·734 grammes (11·37 grains) of uric acid in the twenty-four hours ; and during winter, 1·39 grammes (21·54 grains).

The seat of origin of the uric acid in the animal economy is still a question *sub judice*. One thing is, however, certain, namely, that it is not formed by the kidneys ; in fact, I don't believe that the kidneys form any of the substances they excrete.

Uric acid, like urea, can be detected in the blood, but whether or not it is formed in the circulation or in the tissues it is exceedingly difficult to say. It has been detected in the brain, liver, lungs, and spleen, and, as it occurs in the spleen in the greatest quantity, many regard that organ as the chief seat of its formation. I may add that, like urea, its source appears to be twofold; first, the disintegration of the nitrogenised tissues; and, secondly, the transformation of the excess of albuminous food.

Urea appears to stand in relation to uric acid as a substance belonging to a higher grade of metamorphosis; for whatever accelerates oxidation increases the amount of the eliminated urea, and diminishes that of the eliminated uric acid; while whatever interferes with oxidation decreases the urea, and increases the uric acid in the urine. It has already been shown that urea can be artificially formed from uric acid out of the body, (e) as well as that uric acid is convertible into urea in the body of the rabbit. It may be here added that Stockris has found that the same thing occurs in the human body. (f)

Although, as before said, uric acid is always present in normal human urine, it is never spontaneously deposited from it in a free state. And this arises from the circumstance that all the uric acid in healthy urine is combined with alkaline bases in the form of soluble salts, urate of soda, urate of potash, and urate of ammonia.

In normal urine these salts are all in a state of solution. But whenever from any cause they chance to be in excess, or the water of the urine to be in diminished quantity, they are deposited in an amorphous state, and as such commonly receive the name of urates or lithates. Bird, Bowman, and mostly all the writers of their time, thought that urates con-

---

(e) Ranke obtained urea from uric acid by simply acting upon it at a high temperature with alkali in presence of a ferment.

(f) In order to ascertain whether uric acid becomes transformed into urea in the human body, Stockris, whilst maintaining a uniform diet, took several doses of urates, carefully determining before and after the experiments the amounts of uric acid and urea in the urine. He concludes from the results obtained, that uric acid becomes metamorphosed into urea within the organism, and the seat of the metamorphosis he believes (from experiments on the liver of animals) to be the liver.— Donders, v. Berlin, vol. ii., part iii., p. 260.

sisted in general of the urate of ammonia. Liebig, Heintz, and Lehmann have, however, shown that they most frequently consist of the urate of soda, which is readily distinguished from the urate of ammonia by its much greater solubility. If, for example, a small quantity of urine containing the deposit be put into a test tube, and a gentle heat applied, as soon as the temperature of the liquid gets up to, or a little beyond that of the human frame (100° F.), the sediment, if consisting of the urate of soda, is re-dissolved, and the urine becomes quite clear. If, on the other hand, the deposit be composed of the urate of ammonia, the liquid requires to be heated to near the boiling point before the sediment entirely disappears. On rising in the morning, if I find a deposit in my urine, I can in general readily tell whether or not it is composed of the urate of soda by observing if it disappears on the addition of the warm, freshly-passed urine.

PATHOLOGY.

Deposits of urates are of a yellow or pink colour, according to the amount and kind of urohæmatin present in the liquid. The higher the colour the more febrile the state of the body. At one time their presence in the urine was regarded as a sign of the crisis of disease, and hence they received the name of critical discharges. Modern investigation has, however, proved that they cannot always be looked upon in this light, as they may also appear under a variety of conditions, which, although abnormal in themselves, are not truly diseased states, according to the usual meaning of the word. Thus, for example, urates may be deposited in the urine after a slight attack of indigestion, the result of over eating or drinking. Great exertion, especially if accompanied by profuse sweating, will also cause them to appear.

Hence we find them present in the urine after a fatiguing walk, a long day's hunting, or even after a ball, or other such occasional amusement, especially if it has been associated with much mental excitement. Hard study, and even a fright will in some persons be followed by a deposit of urate of soda. A sudden change in the mode of life is a very common cause of their appearance, as, for example, confinement in town to country people, or a few days' residence in France to an Englishman unaccustomed to French dishes. Under none of these circumstances can the deposit be said to indicate a state of disease. It does nothing more than denote a temporary abnormal condition of system which may soon disappear without treatment. If, however, the deposit, instead of being merely temporary, lasts for some days, that of itself would denote that something more than a mere ephemeral derangement of the system is present, and may deserve immediate attention. A deposit in the urine is always a sign of something being wrong, and although, as we have seen, it may occur from very trivial causes, whenever it takes place without appreciable cause, in the otherwise apparently healthy, it is a sign not to be disregarded, as, under such circumstances, it is not unfrequently either the forerunner or associate of gravel or stone. Uric acid in some form or other is the commonest ingredient of all calculi, and there is no period of life exempt from them.

Urates are a very common deposit in the course of acute disease, and they even seldom fail to recur at some period or other in the course of chronic affections. It is, however, only in diseases of an acute febrile or inflammatory type that their sudden appearance can be regarded as indicative of a crisis. Their sudden appearance is due to an important change having occurred in the condition of the patient, and in general, though not always, it is a change for the better. Such, for example, is observed to occur in cases of gout and rheumatism where the climax has been reached. So also in pneumonia and pleurisy when re-solution and absorption commence.

Should a patient, not labouring under any febrile or inflammatory affection, be every now and then troubled with a pink deposit in the urine without any assignable cause, it will be found, in almost nine cases out of ten, that he is suffering from some chronic affection of the heart, liver, or spleen, with which is associated a tendency to gravel. In all such cases, therefore, steps should immediately be taken to counteract this disposition by the administration of alkaline tonics. Should there, however, be any counter-indication to the direct alkaline treatment, those acid salts are to be employed which, during their passage through the body, are converted into alkaline carbonates—such, for example, as citrates, tartrates, lactates, and acetates. Every now and then an exceptional case may arise, where a mineral acid tonic is demanded, under such circumstances the above rule may be departed from, and the case treated according to its special requirements.

It not unfrequently happens that there are crystals of free uric acid scattered among an amorphous deposit of urates. This is more frequently the case in the course of chronic than of acute affections, and generally arises from there being an excess of acid in the urine which has combined with some of the alkaline base and set a portion of the uric acid free.

The crystals of uric acid form on the bottom and sides of the vessel as the urine cools, and, if large, they may be readily detected by the naked eye, in consequence of their yellow, red, or brown colour. Uric acid, like urea, when crystallised in urine always takes up a part of the urohæmatine or any other colouring matter that may chance to be present. So that, the paler the urine, the less coloured are the crystals; the darker the urine the more coloured the crystals. In abnormal urines containing blue, black, or saffron-coloured pigment, the uric acid crystals are blue, black, or yellow; as in these variously-coloured specimens now before you. The crystals are easily recognised by the naked eye if the urine be put into a wine-glass and the deposit allowed to fall to the bottom.

Uric acid is not necessarily in excess when it crystallises spontaneously—though that is the general rule. It has just been said that an excess of free acid induces its crystallisation. I may now add that it is also deposited when from any cause the proportion of alkaline base is abnormally diminished. This arises from the circumstance that uric acid is much less soluble than any of its salts, even the urate of ammonia included.

The daily amount of uric acid excreted varies very considerably in disease. It is generally in excess during the course of all fevers (yellow fever excepted), exanthematous affections, and inflammatory diseases. It has been found doubled in typhus (Parkes), greatly increased in small-pox, as well as considerably augmented in pneumonia. In certain diseases crystals of free uric acid are spontaneously deposited in the urine. This is more particularly the case in hepatic, cardiac, and splenic diseases.

In liver affections there is frequently a spontaneous deposit of free uric acid among the amorphous urates, which, as already said, are so common in those affections. This is more particularly observable in cancer of that organ, in which case, too, the uric acid is generally in excess ;—a fact frequently made use of in diagnosis, for in non-malignant hepatic disease, especially towards its latter stages, the uric acid is found to be remarkably diminished. I had once an admirable opportunity of verifying this statement in a fatal case of jaundice, about which there was some difference of opinion as to its cause, some of the Medical attendants thinking it a case of cancer ; others of non-malignant obstruction,—the latter view being greatly supported by the fact of there being no excess of uric acid in the urine. This case I had the opportunity of care-

fully watching to its termination, and, gradually, as the liver
ceased to perform its normal functions, the uric acid diminished,
until it at length totally disappeared from the urine. (g)

Date.	Quantity of urine in 24 hours.		Quantity of uric acid in 24 hours.	
	cc.	oz.	grammes.	grains.
November 4	. 1705 =	55	0·511 =	7·920
December 1	. 1333 =	43	0·266 =	4·123
,, 12	. 1023 =	33	None.	

In leucocythemia a great excess of uric acid is said to occur.
It even occasionally happens that a spontaneous deposit of free
uric acid occurs in the course of chronic bronchitis, emphysema, and phthisis. This is supposed to be in consequence of
the uric acid in the blood being imperfectly oxidised into urea.
· It will be remembered that it was already said that urea to
some extent comes from the direct oxidation of uric acid in
the circulation, and therefore, whatever retards oxidation tends
to diminish the quantity of urea and increase that of uric
acid. Hence it is that in all cases of disease attended with
imperfect arterialisation of the blood the urea is diminished,
and the uric acid increased. There are, of course, exceptions
to this rule; in December, 1859, Mr. Murray sent me some
urine from a case of acute phthisis in which there was both
an excess of urea and of uric acid.

In certain pathological states of the system the quantity of
uric acid excreted is occasionally so great that the simple
addition of nitric acid to the urine causes an instantaneous
deposit to occur ; a deposit which, as Beale justly remarks, is
apt to be mistaken for albumen. On one occasion I obtained
a copious deposit of amorphous uric acid from the urine of a
young man aged 19, suffering from pneumonia of the right
side. It was on the eighth day of the disease, and the first

(g) A full report of this case is given in the author's work on Jaundice
and Diseases of the Liver and Pancreas, p. 71.

of the re-appearance of the chlorides. I have observed the same thing in phthisis, Rees has noticed it in typhoid, Parkes in typhus, and Beale has also obtained these uric acid precipitates in a case of hydatid tumour of the liver, and in one of a man suffering from rheumatic fever.

Free uric acid is in general very readily recognised with the microscope, for although the crystals assume a multitude of forms, each form is well defined.

Fig. 9 shows the forms of uric acid most frequently met with.

FIG. 9.

Common forms of Uric Acid.

The crystals may be well-marked diamond-shaped plates, rounded lozenges, ovals, squares, or rods. They may be perfectly solid or composed of condensed bundles of needle-shaped crystals, which are occasionally free at the ends and cause the crystals to resemble, in some degree, a small-toothed comb or a scrubbing brush.

All uric acid crystals polarise, but the play of colours is

best seen in the thin flat plates. Indeed, it is sometimes necessary to place a piece of selenite beneath the slide containing the crystals in order to get the polarised colours distinctly.

When a difficulty arises regarding the chemical nature of a crystal, in consequence of it having assumed an unusual form, add a drop of caustic potash, and after the crystal has become dissolved, neutralise the solution with a little hydrochloric acid, and most probably the uric acid will re-crystallise in one of the more common forms. I find the colour always a great aid to the diagnosis, for no crystals are so prone to take up the colouring matter of the urine as those of uric acid. Indeed, by far the majority of crystalline substances met with

FIG. 10.

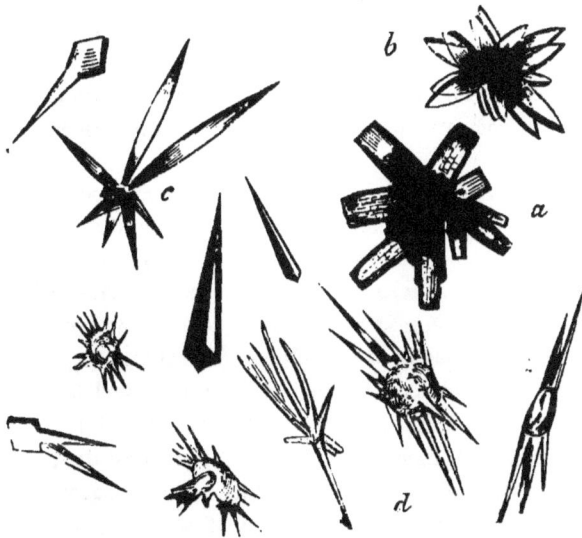

FIG. 10.—Rare forms of uric acid.

in human urine are perfectly colourless, even when crystallised in highly coloured urine.

The subjoined woodcut (Fig. 10) contains all the rare forms of uric acid crystals I have ever met with, and I think I may even venture to affirm that the types of all the various forms of uric acid are represented in these two woodcuts (Figs. 9 and 10).

a and *b* in Fig. 10 are not very uncommon. They are often to be met with in the urine of patients labouring under hepatic disease, more especially those in which there is a tendency to gravel or stone. *c* are rare. They look like old-fashioned arrow-heads. Those represented were sketched from a spontaneous deposit in the urine of a gentleman labouring under diphtheria. On one occasion they had a blueish-purple tint. *d* is the rarest of all the forms of uric acid; I have only once or twice met with it. Those given were sketched for me by Mr. Alexander Bruce, and are most faithful representations of the crystals, the true nature of which was ascertained by chemical analysis. These crystals were spontaneously deposited in the urine of a boy aged 7 years (Hospital case) suffering from albumenoid disease of the liver, spleen, and kidneys.

Both liver and spleen were excessively enlarged, and the urine was highly albuminous. This leads me to remark that the presence of albumen influences to a certain extent the form of crystalline urinary deposits, especially those of uric acid. Crystals of uric acid are frequently deposited round tube casts and masses of epithelium. Sometimes they appear in the form of rows, as if they had been deposited in one of the tubuli uriniferi, and afterwards floated out with the urine. This is not at all improbable, for be it remembered that, under certain pathological conditions which in this country are, unfortunately, far from rare, deposits of uric acid, either free or combined, take place in all parts of the urinary passages — in the urethra, in the bladder, in the ureters, or in the kidneys. The deposit may occur in the form of sand (as seen in this beautiful wax model), as gravel (as represented in this drawing), or as calculi (as seen in this preparation).

Uric acid calculi, although most common in early life, may occur at all ages from birth upwards.

Infants have been born with stone in the bladder. Here are three well-formed calculi, as large as barleycorns, that were found by Mr. Shepherd, of Great Portland-street, in the right kidney of a child aged ten months, about the cause of whose death I was consulted by the coroner in consequence of her having died suddenly after drinking some porter.

This, on the other hand, is the kidney of a gentleman over 60 years of age, and it contains no less than fifty-nine small stones, mostly, as you see, about the size of garden peas. Curiously enough, the right kidney contained none, nor was I able to detect any in the bladder. In both this case and that of the child the calculi consist of uric acid.

Uric acid are the most common of all calculi. Of one hundred and twenty-six of the calculi in the museum of University College analysed by me a few years ago, eighty-seven consisted of uric acid either in whole or in part.

As regards the treatment of uric acid calculi, I may here say a few words.

When the calculus is in the urethra or bladder, cutting or crushing, if the condition of the patient admits of it, ought immediately to be had recourse to. But when the calculus is in the ureter or kidney, it is unfortunately beyond the reach of any instrumental interference; and in the majority of such cases, if the stone be large, incurable. Should, however, the case be seen in the early stage, it is in our power, not only to arrest any further deposit, but even, by suitable remedies, to some extent, re-dissolve what has already formed, more particularly if it be in the shape of sand. When, however, a large solid stone has formed, palliation, not cure, is all that can be hoped for.

The remarks now made are limited entirely to uric acid calculi; for, as you will afterwards see, there are some calculi of so soluble a nature as to admit of removal by dissolution even when in an advanced stage of formation.

Various remedies have been resorted to with the view of retaining any excess of uric acid in a soluble state until after its expulsion from the body. Carbonate of soda has been much used for this purpose, also the bicarbonate of potash, and more recently the carbonate and citrate of lithia. Each of these remedies is useful, as, indeed, all alkaline carbonates are, with the exception of that of ammonia. The carbonate of ammonia should be avoided in the uric acid diathesis on account of the salt which it forms being much less soluble than any of the others.

In treating cases of uric acid deposit, it is unnecessary to render the urine more than neutral. Indeed, there is a danger in making it alkaline from the circumstance that phosphates have a great tendency to form round uric acid calculi (as is seen in these twelve specimens of various shapes and sizes now on the table), and whenever the urine becomes artificially alkaline, although the deposition of uric acid is interrupted, a deposit of triple phosphates takes its place. The stone, therefore goes on forming as rapidly as before, the only difference being that its nature is changed.

In cases where the alkaline carbonates are counter-indicated, citrates, tartrates, or acetates may be used in their stead, for the reasons before given. Some have imagined that uric acid gravel only appears in urine of the well-fed and corpulent. This is a great error. It equally occurs in that of the poorly nourished and emaciated, the only difference being that it is more common among the former than among the latter. In this urine, you perceive, there is a copious deposit of uric acid. The bottom of the glass looks as if it had been sprinkled over with a layer of yellow sand; yet this is the urine of a delicate, anæmic lady, with a miserable appetite. In her case the deposit appears to be hereditary—at least, her father is said to have suffered from gravel.

The most remarkable illustration, however, that I can give of the truth of the above statement is furnished by the sand and gravel in this paper, which was passed by a patient of Mr. Ramsbotham, of Amwell-street. The gentleman who passed it was not only out of condition, but was actually on what might be called starving diet. The following is the brief history of the case, kindly furnished to me by Mr. Ramsbotham :—

" An officer in the army, upwards of 60 years of age, who had seen active service in every quarter of the globe, on his return from India by the overland route, deviated from the usual track for the express purpose of restoring his shattered health in the hydropathic establishment of Prisnitz, in Silesia. He remained there some weeks; but failing to have his juvenility restored, he repaired to a rival establishment a few miles distant, in which a directly opposite mode of treatment was adopted. Instead of using water freely, both internally and externally, as in the other establishment, in the latter very little food and no water was allowed, so long as the patient's strength would permit the deprivation, and then, perhaps, only at distant periods would a wineglassful of water be given. The urine passed was scanty in quantity, and, when simply filtered, left on the paper the powder, of which this is a portion."

On analysis, I found that the powder and small concretions consisted of uric acid, and urate of soda, with a very small quantity of urate of ammonia.

Enough has now been said regarding uric acid in excess ; we shall therefore turn our attention to the equally important fact, that under certain pathological conditions its formation is diminished. We are not yet in a position to speak confidently regarding the nature of all the conditions, but we are sufficiently familiar with the fact that the elimination of uric acid by the urine is diminished in the following diseases, and that it is to some extent under the control of remedies. In yellow (Porcher) and remittent (Parkes) fevers the excretion of uric acid is much decreased. In diabetes it is likewise found to be reduced to a minimum. Lehmann, indeed, thinks that in diabetes it is replaced by hippuric acid; but although I have frequently verified the former, I have as constantly failed to confirm the latter statement of Lehmann's. In the case of saccharine urine from injury to the head (p. 51), during the first few days there was a total absence of uric acid in the man's urine. In albuminuria I have frequently found the uric acid greatly diminished, and absolutely wanting in two cases. It is said also to be deficient in the urine of cholera. In chlorosis, anæmia, and hysteria it is likewise decreased. The excretion of uric acid is diminished before the outbreak of the paroxysm of gout, but in this case it is known to accumulate in the blood. Dr. Garrod has frequently obtained crystals of uric acid from the serum of the blood of gouty patients. His method is to take from one to two fluid-drachms of the serum of blood, and put it into a flattened glass dish or capsule, about three inches in diameter, and one-third of an inch in depth; to this is added ordinary strong acetic acid, in the proportion of six minims to each fluid-drachm of serum, which usually causes the evolution of a few bubbles of gas. When the fluids are well mixed, is introduced a very fine thread, consisting of from one to three ultimate fibres, about an inch in length, from a piece of unwashed huckaback or other linen fabric, which should be depressed by means of a small rod. The glass is then put aside in a moderately warm place, until the serum is quite set and almost dry; the mantel-piece in a room of the ordinary temperature answers very well, the time varying from twenty-four to forty-eight hours, depending on the warmth and dryness of the atmosphere.

In normal blood the acid occurs in such small quantity that it is scarcely possible to detect it.

Uric acid has also been obtained from the serum of a blister placed on a gouty patient, but not when it has been placed over the goutily inflamed part (Garrod). Both before and during the attack of gout the uric acid is diminished in the urine, but no sooner does convalescence begin than it rises even beyond the normal standard. From the twenty-four hours' urine of a convalescent patient, who was under Dr. Parkes in the University College Hospital, I obtained no less than 1·15 grammes (17·825 grains) of uric acid—double the amount the man would probably have passed under similar circumstances had the uric acid not been retained in his system before and during the paroxysm.

During the last two years I have had, at intervals of about three or four months, a patient, aged 65, under treatment, who has been a martyr to very severe paroxysms of gout for, at least, twenty years. All the joints of his hands and feet are distorted from deposits, and he has no less than four well-marked concretions of the urate of soda in his right ear. This man's urine has been carefully analysed on several occasions, and the following is a sample of the results. The attack I now allude to began on November 15, 1863, was at its height on the 21st, and had disappeared on December 10.

Twenty-four Hours' Urine.

	November 18, Third Day.	November 21, Climax.	December 7, nearly Convalescent.
Quantity.	2000 c.c.	1062 c.c.	1620 c.c.
Specific gravity	1011	1010	1008
Uric acid.	0·095	0·023	0·178

On the third day of the attack he came to the Hospital, which was November 18, and was put upon the following mixture :—℞. Vini colchici, ʒss.; liq. potassæ, ♏vi.; Aquæ, ʒss., M., ter in die.

It would appear, then, as immediately after the gouty paroxysm has passed the uric acid increases in the urine, the increase of that ingredient in the course of the paroxysm ought to be considered as a favourable symptom.

The gouty deposits commonly called chalk stones, and which are so frequently met with in the ball of the great toe, in the joints of the fingers, and in the lobes of the ears (of men), in all cases of long standing, are not chalk at all, but consist of crystals of urate of soda (Garrod). As can be readily ascertained by picking out, with the point of the lancet, a minute portion from under the skin, and examining it with the microscope. If a sufficiently large piece can be obtained, it may be tested for uric acid by means of heat, nitric acid, and ammonia, in the manner described at page 41.

A deposit of urate of soda in the cartilages of the joints precedes or accompanies every attack of gout. Hence arises the necessity of freeing the system of this product. Some persons have thought, that as colchicum—the favourite remedy for gout—instead of increasing the excretion of uric acid, diminishes it, and not it alone, but even the urea, the benefits derived from this agent must, if they are not entirely imaginary, be due to some different cause than its action upon uric acid. I believe, however, that colchicum does act upon the uric acid, but not, as is usually imagined, by merely affecting its excretion. On the contrary, it appears to me that by some means or other it arrests its formation, and that it is on this account that an occasional dose of colchicum during the intervals is found to ward off the attack of gout.

Dr. Ranke has shown that the formation of uric acid is diminished by quinine, and Dr. G. Pringle, while working in my laboratory, made some experiments upon himself, the results of which are quite confirmatory of Dr. Ranke's views.

January, 1858. G. P., age 24, healthy and well developed.

Twenty-four Hours' Urine.

Diet.	Quantity of urine.	Sp. gr.	Reaction.	Quantity of uric acid.
	c.c.			grammes. grains.
Mixed ; drink only water, neither tea nor coffee .	656	1031	acid.	1·39 = 21·545
After taking during five days 10 grains of quinine daily ; food and drink as before . . .	665		,,	0·724 = 11·222
The same after seven days of quinine treatment .	602		,,	0·577 = 8·945

The effect of the quinine did not pass off immediately.

Diet as before ; quinine discontinued for two days	600	1036		0·66 = 10·230
Do., do., do., three days .	525	1033	,,	0·792 = 11·270

Quinine combined with colchicum was Bequerel's favourite prescription for gout. Curiously enough, too, he used to combine it with digitalis, another substance which has the effect of diminishing the uric acid. A good formula is—℞ Ext. colchici, gr. x. ; digitalini, gr. ½ ; quinæ sulphatis, Əj ; conf. rosæ, q. s. Fiat massa et divide in pilulas, xij. ; sumat unam ter in die.

While in Munich, a few years ago, when talking to Professor Pfeuffer regarding his treatment of gout, he told me that he found the acetate of potash a good remedy, and it so happens that this also has the effect of diminishing the amount of urea and uric acid in the urine ; and, as I said with reference to colchicum, I believe it is not simply the elimination of these substances which is arrested, but their actual formation that is checked.

I have already pointed out how animal diet increases uric acid. I may mention that there are certain other remedies besides those already spoken of, which diminish the formation of uric acid ; these are atropine and cod-liver oil. Tea and coffee produce the same effect. On the other hand, there are some which, although they do not diminish its formation, yet lessen the quantity in the system by increasing its elimination from the body. Among these may be mentioned the phosphate of soda, and the liquor and bicarbonate of potash. Both in the urine of gout and of rheumatism the uric acid has been found increased after the administration of liquor potassæ, as well as after a few doses of the simple carbonate of potash, and it is not at all improbable that the beneficial action of these remedies in the above-named diseases may in a measure be due to their hastening the neutralisation and elimination of the acid. Indeed, were I asked to give an idea of the pathology of gout and rheumatism, I would say that all our knowledge as yet tends in one direction—namely, to show that both gout and rheumatism are due to the presence of an abnormal amount of acid in the system ; most probably uric acid in the one case, lactic acid in the other. I have repeated Richardson's experiments of injecting lactic acid under the skin of dogs, and although not always successful, have on three or four occasions produced artificial articular rheumatism. On one occasion I even got the endocarditis he has so ably described.

Uric acid is not only eliminated by the kidneys, but by the skin, and even by the lungs ; so that, by increasing the action of the skin and pulmonary exhalation, we may aid its elimination from the blood. Experience has shown that whatever interferes with cutaneous transpiration augments the quantity of uric acid in the urine. The effect of cold has been already alluded to. Skin diseases have a similar effect. In eczema and psoriasis, uric acid increases in the urine, and in the pustules of pemphigus crystals of it have been detected (Malmeston). In the urine of a severe case of pemphigus, under the care of Dr. Hughlings Jackson, which he sent to me to examine, I found 17 grains of uric acid to the 32 oz. The urine was of a dirty brown colour, with a very acid reaction, and had a specific gravity of 1023.

With regard to the influence of alcoholic drinks, beer included, it is interesting to observe how they lessen the urea and increase the uric acid in the urine ; whereas, with tea and coffee, the reverse occurs : they lessen the uric acid and increase the urea. Then, again, there are substances which act similarly on both uric acid and urea. The acetate of potash, colchicum, quinine, and cod-liver oil, for example, diminish, while bicarbonate of potash and liquor potassæ increase both.

These facts furnish us with several hints which may be turned to good account in practice. Remedies we possess in abundance, if we only knew how to apply them.

When I hear men saying that the medicines they are employing are little better than useless, it often occurs to me that they are, perhaps, not far wrong, although the want of success may not be, as they "vainly" imagine, due to the impotence of remedies, but to the want of knowledge and judgment brought to bear on the employment of them. A man needs only to pay a little attention to the teachings of physiology to learn what powerful weapons nature has put into his hands. Like all weapons, however, they may be turned to good or to evil account, according to the knowledge or the inclination of their employer. I would even venture to suggest to our sceptical brethren that it would, perhaps, be wiser if, before passing judgment on any particular remedy, they first made themselves familiar with the how, when, and why to employ it.

LECTURE V.

.

HIPPURIC ACID, IN HEALTH AND IN DISEASE.—CHLORIDE OF
SODIUM : ITS USES IN THE ANIMAL BODY, THE AMOUNT
IN THE URINE DURING DISEASE, AND ITS EFFECTS AS A
REMEDY.

HIPPURIC ACID $C_{18}H_9NO_6$.

CHEMISTRY.

HIPPURIC acid is a beautiful white, crystalline, slightly acid
body. The crystals are, in general, rhombic prisms, columns,
or long fine needles. Occasionally they are massed in dove-
tail-shaped fans. It is soluble in alcohol, ether, and water.
Its aqueous solution is sufficiently acid to redden litmus.
Hence it is a much stronger acid than uric acid, and possibly
it greatly contributes to produce the acidity of normal urine.

The only other urinary sediment which is at all likely to
be confounded with hippuric acid is that of the triple phosphates. The crystals of hippuric acid, however, are easily
distinguished from those of the latter substance by their
occurring in acid urines ; while triple phosphates are only to
be found in alkaline, or at least neutral liquids. In cases of
doubt all that is necessary is to add a drop of hydrochloric

acid, which dissolves the phosphates and leaves the crystals of hippuric acid intact.

When directly crystallised from human urine, hippuric acid is dark coloured, in consequence of its combining with the urohæmatin and other pigments present in the excretion. It can, however, be readily obtained pure by repeated crystallisation, or by passing its solution through charcoal.

By the process of fermentation, hippuric is transformed into benzoic acid, which crystallises in flat, white, shining plates. When urine decomposes the hippuric acid is thus changed. Hence the difficulty of finding it in stale urine.

If a few dry crystals of hippuric acid be heated in a test-tube, they partly sublime into benzoic acid, which collects on the cold portion of the tube, and partly form nitro-benzole, a red liquid with a fine prussic acid-like fragrance, resembling that of bitter almond oil. This odour is readily obtained from a very few crystals, and affords an excellent means of ascertaining the presence of hippuric acid.

Benzoic acid contains no nitrogen, whereas hippuric acid does, as can be easily shown by heating some, along with a mixture of caustic soda and lime, in a test-tube, when the nitrogen is given off in the form of ammonia, which can be recognised by its smell and by its turning white paper, moistened with a solution of the subnitrate of mercury, black. I have particularly mentioned the fact of hippuric acid being at once transformed into benzoic acid (out of the body) by the mere application of heat, because, when benzoic acid is taken internally, either as a remedy or for experiment, the animal body reconverts it into hippuric acid. Hence the tissues or blood, or some other part of the frame, must possess the power of yielding up the requisite quantity of nitrogen, to transform the one substance into the other.

Method of Analysis. — Evaporate rapidly, with constant stirring, from 300 to 600 c. c. (from eighteen to twenty ounces) of urine to nearly dryness. When cold, filter and add to the syrupy liquid from ten to twenty drops of concentrated hydrochloric acid. Stir the whole, and then extract the hippuric acid with pure ether, to which ten per cent. of absolute alcohol has been added. On slowly evaporating the etherial solution, the crystals of hippuric acid are deposited.

If a urinary sediment consists of both uric and hippuric acids, and it be deemed desirable to separate them, all that is necessary is to boil the sediment in alcohol slightly acidulated with hydrochloric acid, and filter the solution. The hippuric acid goes through, and may be obtained from the filtrate by simple evaporation. The uric acid remains on the filter.

PHYSIOLOGY.

Although Liebig discovered hippuric acid in human as well as in herbivorous animals' urine, it was for a long time thought, as it occurred in such small quantity in the human urine, that its presence there might be regarded as accidental. The recent researches of Weissmann and of Hallwachs (made independently of each other) have clearly shown, however, that normal human urine contains a much larger quantity of hippuric acid than was formerly imagined. In fact, they have found that we pass more hippuric than uric acid, and the reason why we remained so long ignorant of this important fact is no doubt in consequence of hippuric acid being so exceedingly soluble in water, that it is rarely, if ever, to be found as a spontaneous deposit.

Hippuric acid, until quite recently, was in general regarded as the crystalline organic acid characteristic of vegetable diet. A view which can no longer be entertained, since Weissmann has shown that its presence in human urine does not solely depend on the kind of food, although, as might be expected, the diet has an important influence on its quantity.

The following table gives the result of his analyses : —

Diet		Hippuric Acid in the Twenty-four Hours' Urine.
Mixed . . .	2·17 grammes	= 33·635 grains
Animal. . .	0·765 ,,	= 11·857 ,,
Bread and water .	0·65 ,,	= 10·075 ,,

The largest amount of hippuric acid passed in the twenty-four hours is always found to follow a purely vegetable diet, although not entirely dependent on it. Certain fruits have a very marked effect in increasing the amount of hippuric acid in the human urine. Lücke specially calls attention to this fact. Cranberries and blackberries he found greatly increased it ; and more recently, Thudichum and others have made a similar statement with regard to plums.

As was before hinted, it has been shown that the hippuric acid found in the urine does not all come from the vegetable food. The researches of Weissmann have proved, as seen in the above table, that on an animal diet more hippuric acid is passed than on a simple bread-and-water diet. This opinion is still further strengthened by the fact that Dessaignes has succeeded in forming hippuric acid artificially. Moreover, benzoic acid, which can be obtained by the action of nitric acid on protein substances, is, as was first pointed out by Mr. Ure, transformed in the human body into hippuric acid. Any one can readily convince himself of the correctness of this last statement by taking ten or twelve grains of benzoic acid at bed-time, and looking for hippuric acid in the morning's urine. In order to be successful in the search, all that is required is to add a few drops of hydrochloric acid to the urine, and then concentrate a little of it to the consistency of a syrup. While cooling, fine crystals of hippuric acid form in the vessel ; and they can be easily detected with the microscope.

The subjoined figure represents the various forms of crystal usually met with when the acid has been directly crystallised from urine. The fine needle-shaped ones, either separate or combined in stellate groups, are the most common. Sometimes they form perfect rosettes.

FIG. 11.

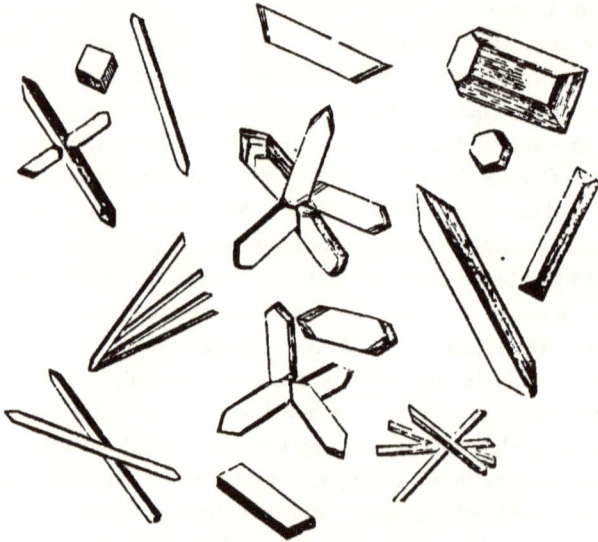

Hippuric acid from human urine.

Maack found that horses kept standing in the stable pass very little hippuric acid, but large quantities of urea, while I made an exactly similar observation regarding our London stall-fed cows. Exercise, therefore, seems to increase the amount of hippuric acid and diminish that of the urea excreted by herbivorous animals.

Hippuric acid, like urea and uric acid, is a normal constituent of the healthy blood of the ox (Verdeil), and I have little doubt, on careful analysis, it will be found in the blood of all other animals in whose urine it normally occurs.

PATHOLOGY.

As yet we know very little regarding the pathological importance of hippuric acid ; but the following are a few of the best ascertained facts :—

Hervier has detected hippuric acid in diseased human blood, and Schlossberger in the scales taken from the human skin in a case of ichthyosis.

In diabetes I have found the quantity in the urine almost always increased. Lehmann, indeed, thinks that it replaces the uric acid in that disease, but this I know is incorrect. Although the salient fact remains that in diabetes the hippuric acid is augmented, while the uric acid is diminished in human urine.

In the urine of typhus patients it is increased ; indeed, Dr. Parkes, in generalizing from the experiments of others, says, that as a rule, the hippuric acid is increased in all febrile affections. Hippuric acid appears also to be increased in chorea (Parkes).

There are certain remedies which have a very marked effect on the quantity of hippuric acid eliminated by the urine.

Kletzinsky (Canstatt, 1859) finds that after taking benzoic acid internally the urea is diminished to 2·5 grammes, or 38·75 grains in twenty-four hours, and the hippuric acid increased to the enormous quantity of 9·5 grammes, or 147·25 grains. Consequently there is only 0·3 grammes less nitrogen excreted than under normal circumstances.

The balsam óf Peru when taken internally has also a marked effect in augmenting the excretion of hippuric acid.

On the other hand, when hippuric acid is taken internally, it passes into the urine unchanged. Kühne has made the interesting observation, that benzoic acid, when given to jaundiced patients, is not transformed into hippuric acid, and concludes therefrom that the seat of the transformation must be in the liver, but further researches are much wanted on this point, as indeed upon all others in connexion with the amount and relations of hippuric acid in the human urine.

CHLORIDE OF SODIUM, CL. NA. COMMON SALT.

The physical properties of chloride of sodium need no description, so I shall at once enter upon its

CHEMISTRY.

The presence or absence of chlorides from the urine is very readily ascertained in the following manner:—

To a little urine in a test tube, add a drop or two of nitric acid, and then a few drops of a solution of nitrate of silver; the immediate formation of a white precipitate indicates the presence of chlorides. If only traces of chlorides are present the urine merely becomes milky. The precipitate may be still further proved to consist of chloride of silver by its being soluble in ammonia and re-precipitated on the addition of nitric acid.

Quantitative Analysis.—The amount of chlorides in the urine is best estimated with a standard solution of the nitrate of silver. The solution is prepared in the following manner:—

1st. Dissolve 29·063 grammes (450·476 grains) of chemically pure fused nitrate of silver in 1000 cubic centimeters (32·23 oz.) of distilled water. 1 c. c. of this solution is equal to 0·01 gramme of chloride of sodium and to 0·006 gramme of hydrochloric acid.

2ndly. Prepare a cold saturated solution of neutral chromate of potash.

Being provided with these solutions, employ the same apparatus as was used in estimating the urea (p. 59), and proceed as follows:—

Take 50 c. c. of urine, and if necessary, render them neutral by the addition of a drop or two of carbonate of soda solution. Then add four or five drops of the saturated solution of the chromate of potash, after which nothing more remains to be done but to pour into the urine the standard solution of nitrate of silver drop by drop until the red chromate of silver makes its appearance. As soon as the red colour shows itself the operation is completed, and all that is required is to see how much of the standard solution has been used, and make the calculation accordingly.

If 20 c. c. of standard solution were requisite, and 1 c. c. is, as we have said, equal to 0·006 gramme of hydrochloric acid, the 50 c. c. of urine must have contained 0·12 gramme of hydrochloric acid. Supposing, now, that the patient made 1000 c. c. of urine in the twenty-four hours, it is evident that he must have passed 2·4 grammes of hydrochloric acid, for

$$\frac{1000 \times 0·12}{50.} = 2·4 \text{ grammes.}$$

PHYSIOLOGY.

Common salt plays an important part in the human organism. It is found in every animal fluid and solid, and seems to be quite indispensable to the development and growth of the body. The presence of this substance in the different fluids and tissues is not merely accidental, for it is in a state of chemical combination with the constituents of the flesh and blood. This is proved by the impossibility of separating the salt from many of them by mere washing, and by its being invariably found in their ashes.

It is highly probable that chloride of sodium takes an important share in the metamorphosis of the animal tissues by adding to the solubility of some, and preventing the decomposition of others. For example, it dissolves pure casein, assists in retaining albumen in a liquid state, and has the power of preventing the coagulation of the fibrin of the blood. While these effects may be regarded as the result of chemical action, there are others, no less important, which appear to be dependent upon its physical properties. It has, for instance, a powerful influence upon the colour of the blood, and Scherer thinks this is due to its mechanical relations. The bright and dark tinge of blood, he says, principally depends upon the form of the blood-corpuscles; when they are biconcave the blood is of a bright colour, when biconvex of a dark hue. The form of the corpuscles again depends chiefly on the endosmotic relations existing between their contents and the fluid in which they are suspended, and this is materially influenced by the amount of chloride of sodium present in the circulation.

I cannot venture entirely to endorse this view, for the simple reason with which you are all familiar—namely, that blood changes its colour when shaken with air and carbonic acid alternately, quite independent of the amount of chloride of sodium contained in it.

Although the quantity of chloride of sodium present in the system varies at different times, yet, like all the other important constituents of the frame, it obeys established laws which regulate the amount. In fact, if we examine into the source from whence the body receives its supply, we shall at once find that the quantity present in the blood must be subject to great fluctuations. The salt is very generally distributed over the earth, in the animal, vegetable, and mineral kingdoms, and, consequently, finds its way into our bodies with the food and drink. All civilised nations use salt very largely as a condiment, and some say that the amount so employed is far greater than the wants of the system demand, for they believe that the food and drink furnish the requisite quantity to the organism. There are others, again, who hold an opposite opinion, which they rest upon the fact that certain tribes in the interior of Africa willingly exchange gold-dust for an equal weight of salt, and that whole herds of animals travel miles and miles to drink at the salt springs, thus proving that the system must feel an inordinate want of that substance.

Those who take the opposite view assert that, so far from the system feeling a want of it, there are whole tribes in South America and elsewhere flourishing without even the knowledge of the existence of salt.

This mode of reasoning appears to me most unphilosophical, for each party seems entirely to overlook the source from whence the food receives its supply. Is it not the soil which furnishes the chloride of sodium to the water and to the plants, and from them to the animals we consume? Now, if the soil in a particular district be poor in that material, the water and the vegetables will contain a minimum quantity—perhaps insufficient for the wants of the human body, as well as for the wants of the herbivorous animals. We know very well that mineral substances are distributed unequally over the earth, and it is quite probable that there is plenty of salt in South America, and yet not enough in the interior of Africa ; and thus it is not in the least degree surprising that the negroes in Africa should value salt more than they value gold-dust, and that the cattle, feeling an equal want, but being unacquainted with any system of barter, should travel thousands of miles to supply themselves with a substance which they cannot otherwise obtain. In civilised countries we have not the same facilities of ascertaining whether or not our food contains sufficient salt, for we always add a quantity in order to please our palates. There is one thing that makes me think the soil of this country scarcely contains sufficient, and that is, that oxen fatten better and quicker when salt is added to their food, so long as too great an excess is not given. Yet no one can doubt but that we take far more than we require, for there is always a very large amount present in the urine,—a quantity greater than could result from the elimination of the merely *effête* material; and the more salt we eat, the greater is the amount excreted by the kidneys.

When salt is injected into the blood it is rapidly eliminated from the system by the saliva, the mucous membranes, and by the kidneys, which elimination would not occur so rapidly did the system require it.

The uses of salt in the animal economy must be manifold, for, as before said, it occurs everywhere. One remarkable thing is, that while the chlorine is chiefly combined with soda in the blood and all the different fluids of the body, it is principally united with potash in the muscles.

Lehmann lately advocated the idea that salt plays a most important part in cell development, and supported the view with many ingenious arguments. That it has important uses is evident enough, but to indicate them individually is very difficult. It seems to have an influence, as already said in a previous lecture, on the excretion of urea, for where much salt is eaten an excess of urea appears in the urine. The two substances, in fact, constantly go together, for in the sweat, in the blood, in the humours of the eye, in the flesh, and in the urine they are always found together.(a)

In starvation, where no salt is taken into the system, none is found in the urine, so it seems as if the body felt an absolute want of it, and held back what it had until it received a fresh supply.

During health the amount of chloride of sodium excreted by the kidneys is always less—(Bischoff says $\frac{1}{3}$, Kaupp $\frac{1}{4}$)—

(a) Beale says that the presence of urea in the urine causes common salt, which usually crystallises in cubes, to crystallise in octohedra, and chloride of ammonium, which crystallises in octohedra, to crystallise in cubes.

than what is taken into the stomach; and this is easily enough understood, as a quantity of it is decomposed, the chlorine going to form the hydrochloric acid of the gastric juice; the soda, the glycocholate of soda of the bile. Besides this, a quantity of it must pass away with the fœces, as well as a little be lost by the sweat, and other excretions, all of which contain chloride of sodium. Then, again, in the growing animal a portion must also remain in the system to assist in building up the frame, into the composition of which it so largely enters.

The quantity of chlorine passed in the twenty-four hours' urine is variously given by different observers. Parkes, after comparing the statements of different authors, puts down the mean amount at 8·21 grammes, or 126·76 grains, in the twenty-four hours, but I agree with him in thinking that these figures are too high; about 5 grammes, or 77·5 grains, of chlorine appears to me to be the average amount passed by Englishmen.

PATHOLOGY.

The quantity of chlorine daily excreted varies very much in disease, and in certain affections the variation is attended with such a marked change in the symptoms of the disease that the state of the one may almost be guessed at from a knowledge of the condition of the other.

In certain affections, as first pointed out by Heller, there is not only a diminution, but sometimes an absolute absence of chlorides from the urine. This is more particularly observable in inflammations with exudation, such, for example, as pneumonia. During the course of the pulmonary hepatisation the chlorides may be totally deficient for two or three days together, but no sooner does resolution occur, and the redux crepitation over the diseased part become perceptible, than they are again detectable in the urine. At first it was thought that the diminution of the chlorides in the urine during the course of pneumonia and other acute inflammatory affections might be explained on the ground of the low diet to which the patients are usually in such cases subjected. This has, however, been ascertained not to be the case. Folwarczny found that in the urine of a patient who, for the purpose of experiment, partook of an abundance of salt with his food—salted soup, salted meat, and salted water —no chlorides could be detected during the exudative stage of the disease. The absence of chlorides from the urine in inflammatory affections, therefore, is due to some other cause than their mere diminution in the ingesta, although that diminution also exerts some influence on the result. Beale made the interesting observation, that in cases of pneumonia where no chlorides are to be detected in the urine, the sputa contains them in abundance.

Chlorides are occasionally absent in fevers. Parkes found them totally absent in typhus, greatly diminished in typhoid and in pyæmia, and somewhat reduced in erysipelas. In the urine now before me from a case of acute phthisis (a) they are entirely absent.

The absence of chlorides from the urine may also aid in the diagnosis of cases of heart disease. Folwarczny, while studying the urine in a case of acute articular rheumatism, observed one day that the chlorides suddenly disappeared, while a small quantity of albumen became detectable, and on that day an attack of pericarditis came on.

(a) In this case, curiously enough, while the chlorides were absent, the urea was very greatly diminished. In a case of pneumonia in a lad, aged 19 (Hospital case), the absence of chlorides was immediately followed by an excess of uric acid ; such an excess, too, that the simple addition of nitric acid to the urine produced a copious deposit of amorphous uric acid, which at first sight was mistaken for albumen. The lad had been leeched, and was taking antimony.

Chloride of sodium is greatly diminished in choleraic urine. When it begins to increase in a case of cholera, it is a most favourable symptom, as much or even more reliable than the increase of urea.—(Buhl, quoted by Parkes.)

It is far less common to find the chlorides increased than diminished. Ringer and Parkes found them increased from ten to twenty fold in the cold and hot stages of ague; but, although increased at these times, there was still no total increase in the twenty-four hours' urine.

Chlorides are never found in the urine as a spontaneous deposit, in consequence of the combinations of chlorine with soda, potash, lime, and magnesia—all being exceedingly soluble salts. Hence the presence or absence of chlorides from the urine can only be ascertaained by chemical means, such as those already described.

Chloride of Sodium as a Remedy.—Some have thought that certain diseases might depend on an insufficiency of chlorides in the system, and consequently that the administration of common salt might be attended with beneficial results. Dr. Cotton, who experimented with it at the Consumption Hospital in cases of phthisis(a), came to the following conclusions : 1st. From ʒj. to ʒij. gradually administered seldom produces nausea, thirst, or derangement of the digestive organs; but larger doses cause sickness and vomiting. 2nd. In some cases common salt increases the appetite and acts as a general tonic. 3rd. Its tonic influence in phthisis is equal to that of bitters. 4th. It does not appear to be deficient in the tubercular crasis, or to have any direct action on the disease when fully developed.   .

(a) *Med. Times and Gaz.*, May 28, 1859, p. 548.

LECTURE VI.

UROHÆMATIN, AND THE ABNORMAL PIGMENTS MET WITH IN
WHITE, YELLOW, GREEN, BLUE, AND BLACK URINE—THEIR
CLINICAL SIGNIFICANCE AND TREATMENT.

UROHÆMATIN, the substance which gives to healthy urine its
peculiar tint, is of more importance to the Clinical Physician
than the majority of urinary ingredients; for the quantity
passed in the twenty-four hours is not only an index to the
tear and wear of the tissues, but the best measure we at present possess of the rapidity with which man burns life's lamp.
The amount of all the ingredients of the urine we have hitherto
considered are much more under the influence of the diet than
that of urohæmatin, the quantity of which may be said to be a
tolerably exact measure of the destruction of blood corpuscles.

CHEMISTRY.

Until the last ten or twelve years little was known regarding
the nature of the colouring matter of the urine, and that little
was unfortunately erroneous. Cruickshanks, Vauquelin, and
Chevreul regarded it as a coloured acid ; Vogel thought it was

uric acid, and Prout looked upon it as murexid. Golding Bird, however, pointed out Prout's error by showing that, while hydrochloric acid destroys the colour of murexid, it increases that of the urine. Berzelius also saw Prout's mistake, from urohæmatin being soluble in alcohol while murexid is not. Heller it was who made the first step in the right direction by studying the effects of acids on different urines, and thereby obtaining a variety of colours, which he believed to be due to a substance which he called uroxanthin. This uroxanthin Heller thought consisted of two colouring matters, urrhodin, a red, and uroglaucin, a blue pigment, none of which was he, however, able to separate from the urine. The second step in the right direction was made by Scherer, who was the first that succeeded in isolating anything like the colouring matter of the urine. The substance that Scherer extracted was a brown powder, so closely allied in composition to the colouring matter of the blood that he thought it might be derived from it.

	Urohæmatin.	Blood hæmatin.
C .	. 58·43	70·49
H .	. 5·16	5·76
N .	. 8·83	11·16
O .	. 27·58	12·59

The next step was made by myself in 1851, when, for the first time, it was found possible to extract from human urine its colouring matter in a perfectly pure form. At the same time I made the interesting discovery that not only does the colouring matter of the urine closely resemble the hæmatin of the blood in its appearance, and its properties, but that it even possesses the additional similarity of containing iron, a substance which till then had never been detected in normal urine. At first this discovery was received by Scherer and some others with scepticism, but now even Scherer acknowledges its truth, so that it is unnecessary longer to treat it as a matter of doubt.

Seeing that the colouring matter of the urine so closely resembled the colouring matter of the blood as no longer to leave
a shadow of doubt that the one is the direct derivative of the
other, it was deemed advisable to name it Urohæmatin, a name
which at once met with the approbation of Scherer and others
who were equally with myself interested in the subject.

Properties.—Urohæmatin is a bright red, non-crystallisable,
organic compound. When freshly prepared it somewhat resembles in appearance dark red sealing-wax, and is sufficiently
soft to receive the impress of the nail. After a time it becomes
hard, and somewhat brittle, breaking with a glistening fracture,
but never losing its fine red tint, at least not after twelve
years' exposure to the air. Urohæmatin is highly soluble in
chloroform, alcohol, and ether, and gives to those liquids a
rich, transparent, port wine colour, varying in depth according to the quantity dissolved. It is likewise soluble in fresh
urine, and gives to that liquid not a bright red tint, but the
tint of normal, healthy, high-coloured urine, the depth of
colour being in direct proportion to the quantity of colouring-
matter added. Urohæmatin is, however, insoluble in pure
water, and even in solutions of either chloride of sodium or
chloride of barium. It is also quite insoluble in sulphuric,
nitric, and hydrochloric acids, even the strongest; but soluble
in the caustic alkalies of ammonia, soda, and potash. In
solutions of tartaric and oxalic acids, urohæmatin is quite
insoluble.

In all the above-mentioned characters it bears a close resemblance to the colouring matter of the blood—hæmatin—the
similarity in the two substances being made still more striking
by the fact, already mentioned, of urohæmatin containing
iron; the existence of which is readily demonstrated by burning a small quantity of urohæmatin on a spatula, and dissolving the ash in a weak solution of hydrochloric acid, and testing
it with the ordinary reagents. The addition of a drop or two

of sulpho-cyanide of potassium gives rise to a fine red colour,
while the addition of a similar quantity of ferrocyanide of
potassium gives a precipitate of Prussian blue, thereby estab-
lishing the presence of iron. If any one still doubts the
existence of iron in healthy human urine, it is sufficient to test
the ashes of any urinary residue as above described in order
to dispel the doubt.

Preparation.—Urohæmatin may be obtained from urine in a
pure state in several ways:—1st. By evaporating a large quantity
of urine until it becomes of the consistence of treacle, taking care
to remove the chloride of sodium and other salts as they crystallize. The colouring matter is extracted from this molasses
like syrup by means of alcohol. The deeply-tinged alcoholic
extract is now boiled and treated with slaked lime until it is
completely decolorised ; it is then filtered, and the compound
of lime and colouring matter well washed with water, and
afterwards with ether, to free it from fat, of which there is
invariably a considerable quantity in human urine. The lime
compound when dry is decomposed by hydrochloric acid, and
the liberated colouring matter extracted with boiling alcohol.
The alcoholic solution is now mixed with an equal portion of
ether, and, after being frequently shaken, allowed to stand a
day or two, in order that the ether may dissolve as much as
possible of the pigment. On the addition of water, the ether
charged with the colouring matter separates, and is decanted.
This etherial solution after evaporation is treated with chloroform, which removes the colouring matter from the animal
resin to which it tenaciously adheres ; and on evaporation the
chloroform deposits the pure urohæmatin. When burned,
urohæmatin leaves a small quantity of ash, consisting solely
of the oxide of iron.

2. Scherer's process as modified by me is equally successful, but, unfortunately, also equally troublesome. It is as
follows:—Precipitate the phosphates and sulphates from fresh

urine by means of the nitrate of baryta. Decant, and to the clear ·liquid add basic and neutral acetate of lead until the urine is decolorised.   Collect the lead compound, which contains the colouring matter, dry, and treat it with hydrochloric acid and alcohol, as in the lime process. The other steps are the same in both. For more minute details see the *Pharmaceutical Journal* for November, 1852, the *Würzburg Verhandl.*, Bd. 5, 1854, and *Journ. f. Pract. Chemie*, Bd. 64, p. 264.

Lastly, an etherial solution may be obtained from urine without any previous manipulation at all, except adding strong nitric acid, and heating the mixture to set the colouring matter free, and then agitating it with a large quantity of ether. The ether dissolves the colouring matter, and floats on the surface of the urine. It is then decanted and evaporated to dryness, and the residue treated with chloroform until pure. The loss by this method is so great that it is seldom attempted, except for rough quantitative experiments.

Quantitative Analysis.—Unfortunately, there is as yet no method by which the quantity of urohæmatin excreted in the twenty-four hours can be exactly ascertained ; I hope, however, soon to be able to publish one, founded on the volumetric estimation of the quantity of iron, as I believe that under normal circumstances almost the whole of the iron found in human urine comes in the form of urohæmatin from the blood corpuscles.   Until then, it may be as well, at least for clinical purposes, to adopt the plan I daily follow, namely, of diluting the twenty-four hours' urine with water till it measures sixty ounces (if the quantity is above sixty ounces, it is to be concentrated), and then adding to about two drachms of it in a test tube half a drachm of pure nitric acid, and allowing the mixture to stand for some minutes.   If the quantity of urohæmatin is normal, the mixture will alter but slightly in tint ; whereas if there be an excess it will become pink, red, crimson, or purple, according to the amount present. Heating the mixture hastens the change in colour; but it is better to do the experiment in the cold, and, if necessary, allow plenty of time for the change to take place.

The object of adding the acid is to set the colouring matter free, for without doing so we are no more able to tell how much urohæmatin is present in the urine than to tell how much iodide of potassium is dissolved in water. A pale urine often contains a large amount of urohæmatin; while on the other hand, a high-coloured urine frequently contains very little; so that the addition of acid is essential. It is not necessary to use nitric acid; sulphuric or hydrochloric acid answers the same purpose, only with them the colour is somewhat different. Now, just as the quantity of iodide of potassium present in a glass of water may be roughly guessed at by the quantity of iodine set free on the addition of nitric acid, so may the amount of urohæmatin present in the urine be guessed at by the depth of colour produced on the addition of nitric acid.

In studying the different shades of pigment found in the urine, the best scale to consult is that designed by Professor Vogel, which forms Plate VI. of Neubauer and Vogel's work, translated by Markham for the New Sydenham Society.

PHYSIOLOGY.

Urohæmatin is without doubt an organic substance belonging to the same class as albumen, fibrin, and casein. It differs from these substances, however, in so far as it can be extracted from the urine in the same state as it exists in that liquid; whereas neither albumen, fibrin, nor casein can be obtained separately in the same condition as they exist in the body.

Urohæmatin being normally excreted by the kidneys, like

all other similar products, its quantity fluctuates with the state of the health—the ebb and flow, as it were, of life. In chemical composition we saw that it closely resembles the colouring matter of the blood—so closely, indeed, as to have given rise to the idea that the one is the direct product of the other. Now, as is well known, the colouring matter of the blood, which is treasured up in the red corpuscles, from its wide distribution and unvarying composition, is supposed to play a most essential part in the processes of life.

A blood corpuscle cannot, of course, be expected to last for ever. Its career is as defined as the career of the individual to whom it belongs. It is born, it grows, it performs the functions for which it was called into existence, it dies, and is removed, and others take its place. With every respiratory act, with every pulsation of the heart, thousands of blood corpuscles are destroyed, and thousands reappear. And although it is difficult to say what may be the exact term of life of a blood corpuscle, yet we know that even at most its span can be little more than a few months.

When we inject the blood of an animal with oval corpuscles (a fowl)into the vessels of one with round corpuscles (a dog), we find that before three weeks have passed away all the oval cells have disappeared. It is true that in this case we have, as it were, placed them under abnormal conditions; but if we make allowance for that, and read months instead of weeks, we still cannot spin out the life of a blood corpuscle to more than a quarter of a year; and few will even be inclined to give to it so much as that.

Taking it for granted, then, that blood corpuscles die, and that their *débris* is removed from the circulation ; if uro-hæmatin be regarded as the *débris*, or the product of the colouring matter of the red blood corpuscles, it is evident that the amount of urohæmatin present in the urine, supposing it all to be eliminated by that channel, must, to some extent, afford us an idea of the daily destruction of blood corpuscles. Pathology, as we shall presently see, gives additional support to this theory; meanwhile, however, I must mention that all the colouring matter found in the urine does not come from the above-named source. A small quantity comes directly from the food; a vegetable diet furnishing the greatest quantity, animal food yielding scarcely any. This is easily explained from the well-known fact that almost all vegetables are much richer in colouring matter than the flesh of animals slaughtered by bleeding, as is the case with the majority of those used for domestic purposes. In fact, the respective influence of a vegetable, and an animal diet on the urohæmatin is well illustrated by the amount found in herbivorous and carnivorous animals' urine. It will be recollected that, in the first lecture, I pointed out the striking difference presented by these urines; the former being of a very dark, the latter of a very pale colour, while man's stands between the two.

There is still another source from which comes a yet smaller quantity of the normal colouring matter of the urine—namely, the reabsorbed effete pigment of the bile—urobiliverdin. The amount of this in healthy urine is, however, so insignificant as scarcely to merit even this passing notice. In disease, on the other hand—jaundice, for example—it is occasionally enormous, far outweighing that of the urohæmatin.

PATHOLOGY.

The colour of the urine varies greatly in disease. It may be perfectly white, yellow, brown, red, black, green, or blue, and each of these tints, in the absence of ingesta capable of accidentally producing them, invariably indicates the existence of grave disease.

Normally coloured urine does not, however, exclude the possibility of disease, for the colour of the freshly passed urine is no absolute criterion either of the quantity or the kind of colouring matter it contains. The abnormal like the normal pigments are often combined with some of the óther urinary ingredients in the form of colourless compounds, and it is not until the compound is decomposed, and the pigment set free, that we can take cognizance either of its quantity or its quality. For example, here are three urines :—1st. A pale, almost colourless urine from a healthy infant, aged 18 months. It has a specific gravity of 1018. 2nd. An equally pale, almost colourless urine from a girl, aged 19, suffering from chlorosis. It has also a specific gravity of 1018. 3rd. A dark straw-coloured, but perfectly transparent urine from a healthy man, aged 33. It has likewise a specific gravity of 1018. In fact, these three urines have been purposely selected on account of their having the same specific gravity. To each of these add a quarter of their bulk of strong nitric acid, and bring them to the boiling point. Watch the change. The infant's pale urine is scarcely altered ; the man's dark urine is only slightly deepened in tint; whereas the almost colourless urine of the chlorotic girl has assumed an intensely red hue. What is the cause of this difference? The infant is in the bloom of health—there is no waste of blood corpuscles in it—all the blood discs it possesses are employed in the development of its frame. The man has arrived at maturity ; he is still in the prime of life, and in the enjoyment of perfect health ; his blood corpuscles are not wasted, but merely consumed in the tear and wear of every-day life. The young woman, on the other hand, is suffering from chlorosis ; she has a pale lip, and a blanched cheek ; her corpuscles are being too rapidly consumed ; her life's blood is oozing away by the kidneys, and there it appears as an excess of urohæmatin in her urine.

Take again these two urines so different in appearance. They are from young men about the same age (24 years). The

one urine is perfectly colourless, like water (a) ; the other is
of a deep red colour,—case of hæmaturia from disease of the
kidney. On adding strong hydrochloric acid to the colourless
urine it rapidly assumes a port-wine tint, whereas the same
amount of acid added to the red urine, instead of heightening,
actually destroys the colour it already possesses. And why ?
Simply because the pale urine contains an excess of combined
urohæmatin, which is liberated by the acid ; whereas, the red
urine contains merely a number of free blood corpuscles, which
become coagulated, and, as the colouring matter in them is
insignificant in quantity when compared with the amount of
urohæmatin in the other, no sooner are their cell-walls
destroyed and the contained hæmato-globulin set free and
coagulated, than the red colour disappears. Now, which of
these two classes of urine denotes the most danger? Most
assuredly, not that containing the free blood-cells. A very
small quantity of blood will sometimes colour a great deal of
urine ; whereas an immense destruction of blood corpuscles
may take place in the body, and their *débris* be so eliminated
as to be invisible to the eye until the application of an acid sets
it free. In fact, experience has shown me that the normally
coloured urine of disease is a most treacherous guide to go by.
It often lulls the inexperienced into the belief that there is
nothing materially wrong, when a grave lesion is making rapid
strides towards a fatal termination. Not very long ago I was
told by an intelligent Practitioner that a young lady, regarding
whose health we were consulting, was labouring under hysteria. " The secretions," as he termed them, being " all
right," my opinion had been asked more with the view of
satisfying the friends who were getting fidgety than anything

---

(a) The patient from whom this urine came was sent into the Hospital
(under the care of one of my colleagues) from his being supposed to have
stone. None, however, was detected, and his symptoms appeared to arise
from excessive spinal irritation. At the time the above sample of urine
was passed, the lad had already been six weeks in the Hospital.

else. The case appeared to be what he said, until he added, "She is well fed, and yet she loses flesh, and I don't know why." This remark at once brought us back* to .the urine, which he assured me was perfectly natural in colour, and contained neither sugar nor albumen. I analysed some, and the case proved the very counterpart of the lad's just spoken of. Although the girl's urine was natural in colour, her life's blood was imperceptibly oozing away with it. This is no solitary example ; I might cite many cases of so-called hysteria, which were in reality cases of serious, though obscure, disease ; but that is surely unnecessary. Who amongst us has not seen patients die, and their disease put down as, only hysteria ? The time is, nevertheless, not far distant when we shall learn that hysteria is something more than "mere functional derangement." Hysteria may be a convenient term ; but, after all, it is only a cloak of ample dimensions which hides the rags of ignorance. It is a name instead of an explanation, a sham instead of a reality. And what is still worse, it tends to keep us in ignorance by stifling legitimate inquiry ; for no sooner do the majority of practitioners find a name for the disease than they cease to fathom its cause. In some cases of cerebral and spinal disease the excess of urohæmatin in the urine is so great that after it has been set free by an acid and taken up with ether, the ether, after standing, solidifies into a red-currant jelly-like mass, and may actually, in some cases, be cut with a knife.

The best way of showing this is to boil four ounces of urine ; then add nitric acid to set the colouring matter free. When cool, put the urine into a six-ounce bottle along with an ounce of ether. Cork the bottle, thoroughly shake it, and afterwards place it aside for twenty-four hours. At the end of that time the ether will sometimes be found to be like a red tremulous jelly. Such a case is, of course, a very bad one ; but these are not nearly so uncommon as one would imagine. In some of the worst cases of urohæmaturia the urine is neutral or even alkaline, and the *fons et origo mali* is to be looked for in the brain or spinal cord. Indeed, we may often be led to a correct diagnosis of obscure cerebral and spinal disease, by finding urohæmaturia associated with a saccharine or a phosphatic condition of the urine.

In a case of chronic cerebral disease, which I saw with Mr. T. Carr Jackson, the quantity of urohæmatin regularly varied *pari passu* with the severity of the paroxsyms, nitric acid turning the urine red, purple, or even bluish (blue matter being, after a time, deposited on the bottom and sides of the test tube), according to the severity and duration of the attack. It ought never to be forgotten that alkaline remedies, as well as certain vegetable foods, may increase the urohæmatin. Even the external use of carbolic acid sometimes turns it black.(a)

When urohæmatin exists in a free state, the urine is red in colour before any acid is added. These cases differ from hæmaturia, in the urine being clear and transparent, and devoid of blood corpuscles, and from intermittent hæmaturia in the absence of congestive casts. If it contains a deposit, the deposit may or may not be high coloured ; but in any case the supernatant liquid is clear as well as red. Another fact, which is of great clinical importance, is that the urohæmatin is not always in the same state of oxidation; and, like indigo, its colour depends on the amount of oxygen it contains. It may be yellow, red, or brown. In consequence of this, different acids act upon the urine differently. In one case nitric, sulphuric, or hydrochloric acid may produce the same results ; in another hydrochloric acid turns the urine red, while nitric acid makes it blue, green, or yellow. In a third sulphuric acid may develope the colour better than the others.

To illustrate these remarks—Take the case of a gentleman 33 years of age, whose urine to the eye appeared perfectly normal in colour. On adding strong nitric acid, however,.

(a) Almen on a "Change in the Urine from the External Use of Carbolic Acid."—*Lancet*, October 19th, 1870, to this urine it immediately became of a blood-red hue, whereas hydrochloric acid had no effect upon it until some minutes had elapsed, when it gradually caused it to assume the same tint as that produced by the nitric acid. When sulphuric acid is slowly added so as to fall to the bottom of the test-tube, a reddish-brown line appears at the point of contact, and this gradually deepens until, in the course of six hours or so, it assumes a more intense colour than that produced by either of the other acids.

This is an important case in another point of view—namely, as proving the value of urinary analysis in cases of obscure disease. The only symptoms that this patient labours under are those of an irregular kind of dyspepsia, with an occasional pain in the epigastrium. He is in easy circumstances, takes plenty of nourishing food, wants for nothing, yet he gradually gets weaker and weaker, and has lost 17 lbs. in the course of the last year. Even the locality in which the patient lives is a remarkably healthy one (he was sent to me from Anglesey by Dr. D. Williams, of the Menai Bridge), and if it had not been for the condition of the urine, his symptoms would have remained a mystery. This, too, was one of those cases in which the amount of urohæmatin was so great that it caused the etherial solution to solidify into a jelly.]

An excessive excretion of urohæmatin is not limited to cases such as we have been describing. It occurs to some extent in several diseases, especially those in which there is an excessive tissue metamorphosis, and consequent too rapid blood consumption. Hence we occasionally meet with it in low fevers, in diphtheria, in pneumonia, and some other inflammatory affections, in lesions of the nervous system, during an attack of gout, after the fit of ague, and during convalescence from nearly all grave diseases. It is, however, in chlorosis (either in the male or female), and the many unnameable obscure affections of that class, where it becomes a dangerous symptom. In fact, it always indicates the existence of a past or present mischief meriting the closest attention of the Physician, and where we cannot remove the cause we must at least attempt to check the effects of the symptom—namely, to restore to the blood as much as possible of the material which is being drained from it. Every one knows the effects of iron upon the system, but iron alone is not always sufficient for our purpose. Something more is wanted, and that is best supplied by the preparations called the syrups of phosphate of iron. Many syrups have been sold under this title. There are the American syrups; the compound syrup of the phosphate of iron, or "chemical food," as it is sometimes named; the syrup of the superphosphate of iron and lime (excellent for children); and a few others. One and all of them are good in particular cases, but they must always be associated with a judicious selection of regimen in order that the full benefit may be derived from them.

The preparation of zinc, in grain or grain and a-half doses, are also occasionally useful astringent tonics when the drain is very great.

In some cases of disease the excretion of urohæmatin by the kidneys appears to be diminished; but this is only when the system has been so drained that there is little more to come away. In the last stage of chlorosis a great diminution in the amount of the urohæmatin in the urine takes place; so also in chronic cases of hæmaturia, notwithstanding that the urine perhaps looks red. Likewise in cases of chylous urine, in the albuminuria of pregnancy, and in chronic Bright's disease. In all these cases the blood has already been well drained of its constituents before a marked diminution in the amount of the eliminated urohæmatin takes place.

Blue and Green Urine.—A great sensation was created some years ago among Clinical Physicians and Pathologists when the first cases of blue and green urine were reported; all doubted, and many disbelieved in their existence. The least uncharitable thought that the patient had hoaxed the Doctor. Knowledge has, however, advanced since then. Not one, but many observers have met with the same conditions; and now the doubters and disbelievers in the existence of either blue or green urine are only to be found among the ignorant.

To Dr. Hassall we owe the first good report on the existence of blue urine. It was entitled—" On the Frequent Occurrence of Indigo in Human Urine, and on its Chemical, Physiological, and Pathological Relations." (b)

I must here guard you against falling into the common error of supposing that the urine is blue or green at the moment of being voided, for, as far as I am aware, such a condition has never yet been observed. In the cases hitherto reported it has only been after exposure to the air, or after the application of chemical agents, that the urine has been observed to become of the colours alluded to. In Hassall's best-marked case, for example, the urine was of a light brown colour when voided, somewhat alkaline, and had a specific gravity of 1017. In the course of three or four days it became thick and turbid, deep brown, greenish, bluish green, yellowish green, and finally nearly black; the scum on the surface remaining of a permanent, deep indigo-blue colour. All these changes followed upon simple exposure to the air; but, as just said, chemical agents may produce one or more of these changes. Thus, for example, in Munk's case of green urine, (c) when passed the urine was of a dark red colour, but on the addition of ammonia it became green. Like everything else when properly understood, these changes have nothing mysterious about them. The body has not created anything new in making blue or green urine. Indeed, as in disease no new functions

(b) *Philosophical Transactions*, p. 297, 1864.
(c) *Allg. Med. Central Ztg.*, No. 26.

are ever created, but only a change occurs in the rhythm and
force of those already existing, so in disease no new substances
are ever created, but only a change takes place in the quantity
or in the quality of those normally existing. It is now twelve
years since I adopted these views, and the longer and deeper
I peer into the wondrous workings of the human frame, the
more am I satisfied with them, and the easier do I find the
comprehension of disease. Physiology and Pathology are but
one study; and although a man may be a Physiologist without
being a Physician, no man can be a Physician without being a
Physiologist. The same laws that regulate health regulate
disease. The very effects we are now studying are a striking
illustration of the justice of these observations.

For example, the discovery of indigo in urine was thought
to be very extraordinary,—a vegetable product manufactured in
an animal body! The thing appeared absurd; and, as I before
said, the least uncharitable thought the patient had hoaxed
the Doctor. The further science advances, however, the
nearer do we see the animal and vegetable kingdoms approaching each other, and the less surprised are we at finding that
the products of the one are but modifications of the products
of the other. Fat can be got from plants as well as from
animals; sugar can be got from animals as well as from plants.
We have albumen in the plant, as we have albumen in
the hen's egg; we have casein in the milk of the cow,
and we have casein also in the plant. The Chinese know
these things apparently even better than we do. When
they have no milk and want cheese, they make it from
common peas by boiling them along with lime. The cheese
called Tao-fo is the product of the garden, not of the stall.
So, again, with colouring matters; each animal colouring
matter has its counterpart in the vegetable world. The
green colouring matter of the bile is the analogue of the
chlorophyll of the leaf. The red colouring matter of the blood

is similar to the blood red of the rose. Blue urohæmatin is the counterpart of indigo from the plant. The analogy does not only hold good in colour; for, as is well known to chemists, there is also a great resemblance in the composition of the different animal and vegetable pigments, as regards the amount of H, O, N, and C they contain. Moreover, in 1852-53 I traced a still further similarity in the composition of biliverdin and chlorophyll, hæmatin and draconin, urohæmatin and indigo, by finding that they all possessed the common character of containing iron. And what is still more, I believe the day is not far distant when the chemist will copy the Chinaman with his Tao-fo, and resolve those vegetable and animal pigments into one another by adding or subtracting a small quantity of oxygen.

It is well known to everybody that indigo is originally not a blue, but a white substance, and it is only in the process of manufacture that it changes its colour from white to yellow, yellow to green, and green to blue; and there is even another grade of oxidation in which the blue indigo is transformed, to red. Now, as with this vegetable pigment, so it is with the animal pigments; they are nothing more than grades of oxidation of a white radical, as I first hinted in the papers on urohæmatin, already spoken of. If we adopt this view, we can readily understand how easy it is for the red hæmatin of the blood to be changed into the green biliverdin of the bile.

All known pathological data support this view, whilst all the known physiological data can prove nothing to the contrary.

Hitherto I have only spoken of blue and green urine; but there are such things also as blue and green pus(d)—nay, more, both Chevreul and Lecanu have during disease(e) met with blue colouring matters in the blood itself.

(d) I have met with one case, and several others have been published. See " Robin and Verdeil," vol. iii., p. 492. Also Herepath, in " Med. Times and Gaz.," Sept., 1864.

(e) *Ibid*, p. 572. Vide also Carter, in Ed. Med. Journ., 1859, p. 119.

This is not the place to discuss the subject, or I could soon bring a mass of evidence in support of the theory that all animal colouring matters originally spring from one colourless radical. Indeed, so clear does the thing appear to my own mind that I no longer regard it in the light of a theory, but as a fact. A few minutes ago it was said that the transition stage, from white to green and from green to blue indigo, was yellow. Well, even this has its counterpart in the animal body. Sigwart, Sanson, Lecanu, and Chevreul have all described a yellow colouring matter which they found in the blood; and who can doubt its existence, for every day we have ample evidence of the fact of red hæmatin being changed into yellow pigment in the cases of jaundice that are constantly being brought before us.

It is unnecessary even to go so far as that to prove the proposition; for let us take the case of the urine itself, which is alone sufficient to prove it. In case I might be regarded in the light of a one-sided advocate, I shall now quote the observations of other authors, and in order to make the ground more secure, shall select a German, French, and English reporter.

In Munk's case, already referred to, the observations were made upon a man aged 39, while in Hospital, under the care of Professor Traube. The patient stated that he had enjoyed good health until about two years before his admission, when he began to suffer from pains in the loins and trembling of the lower limbs. These increased, and he became so weak that he had to give up work. The urine was found (1) to be of a dark red colour; (2) to contain neither albumen nor bile-pigment; (3) to stain filter-paper red; (4) to become darker when heated with nitric acid; (5) to assume with sulphuric acid a fine deep violet colour; (6) to become decolorised when hydrochloric acid and zinc were added to it; the red colour could, however, be restored by adding nitric acid; (7), and lastly, with ammonia, the urine became green. The quantity passed in twenty-four hours was much below the average; it varied from about fourteen to twenty-three ounces. A very small amount of chlorides and urea was found in the urine.

Munk separated the colouring matter from the urine in a pure state, and found that it was insoluble in water, but soluble in alcohol, ether, and chloroform. It dissolved in soda solution with a pale red colour; in ammonia, with a green; in hydrochloric acid, with a red; in sulphuric acid, with a dark violet hue; and in nitric acid, it became colourless. When burned, it left an ash containing iron.

In Braconnot's case, cited by Robin and Verdeil, vol. iii., p. 491 of their "Chemie Anatomique," the urine was of an obscure yellow colour, and, after standing, the sediment which was deposited from it became of an azure blue. When separated, this blue substance gave to hot water a brown tint, which became of a fine red ("beau rouge") on the addition of strong acids. Moreover, the hot alcoholic solution, which was of a green colour, yielded a blue precipitate on becoming cold.

If further evidence be wanting to prove that the various coloured urine pigments are but different grades of oxidation of urohæmatin, we have it in Hassall's case. At page 298 of the paper already cited it is stated that "The first urine of the series, when passed, was somewhat alkaline to testpaper, had a specific gravity of 1017, and was of a light-brown colour; there formed on its surface, in the course of three or four days, a thick, greasy-looking, soft scum, consisting of vibriones, and very many large and fine crystals of triple phosphate; at about this time the scum became greyish-blue, lavender, bright blue, and finally, after three or four days more, of a deep indigo-blue colour, which was permanent."

Thus are explained the various colours and reactions of the urine that have been recorded by Gilchrist, Heller, Vogel,

Schunck, Lindsay, and a number of others, who have, within
the last few years, so enriched the literature of the subject.
All these changes which occur in urohæmatin out of the body,
as the effects of reagents or simple exposure to the air, are
primarily due to its constitution having been altered in the body
by the effects of disease. I have seen in diphtheria, in low
fever, in phthisis, and in many other exhausting diseases where
the vital powers were much reduced, and the arterialisation of
the blood interfered with, hydrochloric acid throw down blue
crystals of uric acid. Although there is sufficient blue urohæmatin ("uroindigo") to colour the uric acid crystals, there
is often in these cases not sufficient to render the entire urine
blue. Occasionally uric acid is deposited of a purple colour. In
all such cases it will be found that whatever supports the vital
powers tends to diminish the amount of uroindigo in the urine.

I must here warn beginners against falling into the error of
mistaking the colours of the urine produced by remedies for
those arising from disease. Thus santonine, picrotoxine, gam-
boge, rhubarb, and senna all cause the urine to become yellow ;
while turpentine turns it violet, and arsenuretted hydrogen
and creosote make it black. So that one must always begin
by asking what remedies the patient has been taking.

Most coloured remedies produce no effect on the colour of
the urine.

Brown and Black Urine.—There are two entirely different
diseased states in which brown and black urine are met with.
In the one, the urine at the moment of being passed is of a
deep straw colour, and perfectly transparent ; but becomes
gradually darker and darker in tint (without, however, losing
its transparency) until it is as black as porter. In the other, the
urine varies from a chocolate brown to a brownish-black hue
at the moment of being passed. It is not usually transparent,
and deposits a quantity of flocculent sediment, which, when
examined with the microscope, is seen to be made up of brown

or reddish-brown granular pigment, a few blood discs, and frequently a quantity of renal tube casts, together with crystals of uric acid, etc. In this last class of cases the dark colour of the urine is not due to an excessive metamorphosis of blood cells, and consequent increased elimination of abnormal uro-hæmatin, but to a local disorganisation of blood corpuscles, either in the kidney alone, or in the liver and kidney combined; but as the consideration of this condition more properly belongs to the chapter on hæmaturia, we must leave it for the present, and merely direct attention to that peculiar condition of the urine which is supposed to result from the presence of melanin.

In 1858, Dr. Eiselt, of Prague, called attention to the fact that in certain cases of disease the urine, although quite clear when passed, gradually becomes of a dark colour, in consequence of a process of slow oxidation going on. He further showed that by the employment of an active oxidising agent, such as nitric or chromic acid, the freshly-passed urine could be instantly made to assume a black hue. Moreover, he observed that in the three cases of this kind which fell under his observation, the patients all laboured under melanotic cancer, and I have myself added another to the list.(f). The presence, therefore, of melanin in the urine in obscure cases may occasionally help us to a correct diagnosis. It is, however, necessary to remark that a somewhat similar state of the urine, as here described, is occasionally met with in other examples of disease, besides that of melanotic cancer. Great care, therefore, must be taken not to confound them; for although black urine invariably indicates the existence of serious mischief, all the cases in which it occurs, either after the addition of reagents or of simple exposure to the air, are not necessarily hopeless; for just as we may have a green

(f) "On Jaundice and Diseases of the Liver and Pancreas," p. 68.

or a blue urohæmatin, so we may have a brown or a black one. There is nothing incomprehensible in this when we consider that melanin, like the other animal pigments, is a derivative of the colouring matter of the blood, and like it contains iron.

While working in Robin and Verdeil's laboratory in 1852, I prepared a quantity of pure melanin (g) from a melanotic cancer, and compared it with the brown pigments normally existing in the body—among others, with that of the choroid coat of the eye—and I satisfied myself that the normal and pathological, when perfectly pure, are not distinguishable from each other, either by their appearance or properties.

Hence one can readily understand why we should occasionally think we have got melanin in the urine in cases where no melanotic cancer exists. In the article on jaundice already referred to, a case is related of the kind which fell under my notice in 1850. The patient was a woman, aged 22, suffering from an attack of jaundice ; the urine was slightly albuminous, and became black on being boiled with nitric acid. As she perfectly recovered in the space of six weeks, it is not at all probable that she laboured under malignant disease. Since then, three other cases of suspicious urine of this kind have fallen under my notice, and in all of them the urine was, as in the above, albuminous.

Although, then, it is no doubt quite possible to have albumen in the urine in a case of melanotic cancer, yet, as albumen is an almost invariable concomitant of the brown and black urine of other diseases, while it is only exceptionally present in that of malignant affections, we may perhaps, as I shall afterwards show, be able to make use of its presence as a distinguishing sign, especially if it be in accordance with the history and other symptoms of the patient.

(g) " Chemie Anatomique," vol. iii., page 394.

LECTURE VII.

PHOSPHORIC ACID, AND PHOSPHATES IN DISEASES OF THE
  SPINAL CORD, BRAIN, BLADDER, RICKETS, AND OTHER
  AFFECTIONS; PHOSPHATIC GRAVEL AND CALCULI, THEIR
  DIAGNOSIS AND TREATMENT.

ALTHOUGH not venturing to endorse the assertion of a recent
writer, that "a man with too little phosphorus in his brain
is an imbecile, and one with too much a maniac," we
cannot but admit that phosphorus, in some form or other,
is an indispensable ingredient of the human frame. This
arises from the fact that no other element with which we are
at present acquainted seems capable of supplying the place of
phosphorus in the animal tissues.

In a free state phosphorus is never to be met with, either in
the body, or in any of the excretions, except when it has
been taken internally in medicinal or poisonous doses. In a
combined state, on the other hand, it enters very largely into
the composition of our frames, and appears to perform a very
multifarious part in the processes of life. Not only is phosphorus to be found in human bones, and in human brains,
but even in the bodies of the minutest animals and plants in
which all differentiation of tissue and of function is alike
impossible.

CHEMISTRY.

The element phosphorus is, as is well known, a yellow, wax-like solid, easily melted, and highly inflammable. It takes fire spontaneously when raised to a temperature about equal to that of human blood, and has, consequently, to be kept under water. In a free state phosphorus cannot be said to exist in Nature, but in an oxidised form, as phosphoric acid united with lime or some other base, it is widely distributed throughout both the organic and inorganic universe; so that the human body may be said to draw its supply, not only from the food we eat and the water we drink, but even from the very air we breathe.

Phosphorus is excessively easily oxidised, and combines with oxygen in the three proportions of one, three, and five (PO, PO_3, PO_5), all of which combinations are acid substances. It is, however, chiefly, if not entirely, with the last that we have to do, and its presence or absence from any fluid can be ascertained in the following manner :—

To the suspected liquid add a little acetate of soda, with a drop or two of acetic acid, then slowly pour in a small quantity of the sesquichloride of iron. The formation of a gelatinous precipitate of the phosphate of iron (FeO,PO_5) at once reveals the presence of the suspected substance. When only a small quantity of phosphoric acid is present, the precipitate is slow in forming, and the liquid at the same time becomes red, in consequence of the action of the acetate of soda on the excess of iron in the mixture.

The phosphoric acid found in the urine is not in a free state, but is combined with the two alkaline bases, potash and soda, and the two earthy bases, lime and magnesia. Normal urine contains about four times more of the alkaline than of the earthy phosphates, while of the earthy phosphates themselves there is about double the quantity of phosphate of magnesia that there is of phosphate of lime. Neubauer, Kerner, Lehmann, Böcker, and Kletzinsky are of an opposite opinion. They think the phosphate of lime preponderates over the phosphate of magnesia. The different results obtained by these observers, although no doubt chiefly due to the differences in the diet of the persons operated upon, may also in some measure arise from the differences in the ages of the patients, for, I think, it will yet be found that while in the urine of the growing child the phosphate of magnesia preponderates, in the urine of the aged it is the phosphate of lime that is in excess. The child's body retaining the lime in order to form its bones and teeth, while that of the aged person, having ceased to require a supply for that purpose, allows it to pass away with the urine.

Phosphates are never deposited spontaneously from healthy urine. No sooner, however, does the urine become stale, and its urea decompose, than crystals of triple phosphate ($2MgO,NH_4O,PO_5$) are precipitated. This arises from the circumstance that the earthy phosphates are only soluble in acid solutions, and when the urea is decomposed it is changed into carbonate of ammonia, which causes the urine to become alkaline. A similar result may be obtained artificially by simply adding a few drops of ammonia to freshly-passed healthy urine; an immediate white precipitate of the earthy phosphates being thereby induced.

The precipitate consists partly of the ammoniaco-magnesian phosphates (triple phosphates) and partly of the phosphate of lime. Phosphates are in general fine microscopic objects, but the appearances they present greatly depend on whether they are rapidly or slowly crystallised. When the ammoniaco-magnesian phosphate crystallises rapidly, as happens on the addition of ammonia to freshly-passed urine, it assumes the form of fine feathers and fronds, either singly or in stellate groups. Fig. 12. Whereas under the same circumstances the

simple phosphate of lime falls down as an amorphous powder,
or assumes the form of round globules.

FIG. 12.

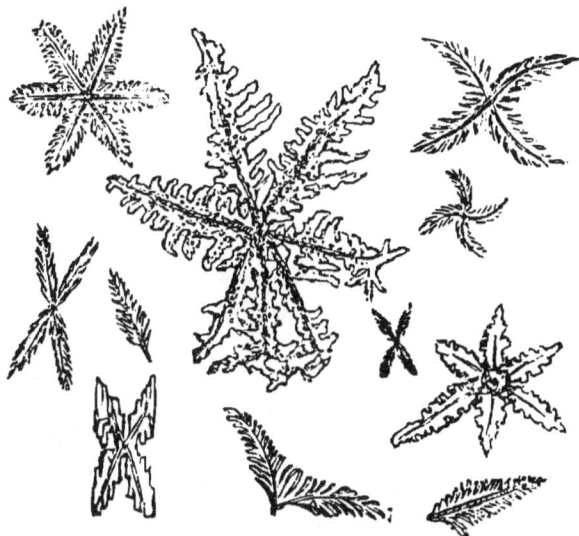

FIG. 12.—Crystals of triple phosphates rapidly deposited in human urine.

On the other hand, when the ammoniaco-magnesian phosphates are allowed slowly to crystallise, as, for example, when
the urine becomes gradually alkaline from the decomposition
of its urea, and the consequent formation of the carbonate of
ammonia, the triple phosphates, instead of appearing as feathers
and fronds, assume the form of well-defined prismatic crystals,
large enough, in most cases, to be visible to the naked eye. They
are usually many-sided prisms, terminating in abrupt angles and
facettes, the largest of the crystals being beautiful polariscopic
objects. A selenite plate adds greatly to their beauty, for by its
means a fine red field and deep blue crystals may be obtained.

It is only the prismatic and modifications of the prismatic
form of crystals that polarise. The feathery phosphates show
no play of colours.

As the prismatic crystals of the triple phosphate frequently form a scum on the surface of urine, the idea has got abroad that all prismatic crystals so found must necessarily be

FIG. 13.

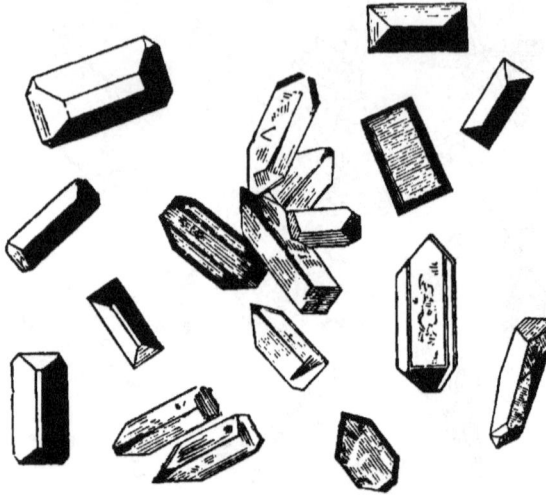

FIG. 13.—Crystals of triple phosphates slowly deposited in human urine.

phosphates. This has led to the invention of many ingenious arguments in order to explain why this scum and these supposed crystals are occasionally met with in acid urine, it being well known that triple phosphates are soluble in acid solutions. The fact of the case, however, is that the prismatic crystals which appear in acid urine, as I shall afterwards have occasion to show, are frequently not phosphates at all, but a crystalline organic substance of an important nature—creatinin.

As regards the simple phosphate of lime ($3CaO,PO_5$), which, when rapidly precipitated, falls down as an amorphous powder, Dr. Hassall has shown that, when allowed to crystallise slowly, as, for example, under the condition just alluded to—the decomposition of urea in normal urine,—it assumes a crystalline form. The crystals of the simple phosphate of lime appear in this case also in the form of prisms. The conclusions arrived at by Hassall are :—

1st. That phosphate of lime occurs in human urine in a crystalline form.

2nd. That it is of frequent occurrence.

3rd. That it is of greater pathological importance than the triple phosphate.

The foregoing remarks are only applicable to the earthy, for the alkaline phosphates are so soluble that even the addition of ammonia does not cause them to crystallise. If, for example, I filter a urine to which ammonia has been added, and thereby separate the earthy phosphates, which of course remain behind on the filter ; on adding to the clear filtrate a little carbonate of ammonia, and some sulphate of magnesia, a second white flocculent precipitate falls down, which, when examined with the microscope, is found, as in the first case, to consist of beautiful feathery crystals of triple phosphates. The explanation of this fact is that the phosphoric acid, which was united with the soda and potash in the form of soluble salts, has, on the addition of the ammonia and magnesia, left the soda and potash, and entered into combination with the ammonia and magnesia in the form of the insoluble ammo-niaco-magnesian phosphate, while the carbonic acid of the added ammonia, and the sulphuric acid of the added magnesia, have combined with the soda and potash in the form of soluble salts, which remain in solution.

It is seen, then, that a quantitative analysis of both the earthy and of the alkaline phosphates in urine might easily be made by proceeding as has just been explained, all that is required being to operate on a given quantity of urine and collect, dry, and weigh the two precipitates. That obtained by the simple addition of ammonia would represent the earthy phosphates of lime and magnesia; while that subsequently obtained by the addition of the carbonate of ammonia and sulphate of magnesia would correspond to the amount of the alkaline phosphates of soda and potash in the secretion.(a)

There are no less than three different forms of phosphate of soda in human urine—an acid ($NaO,PO_5 2HO$), a neutral ($2NaO,PO_5,HO$), and an alkaline ($3NaO,PO_5$).

Some think that the acid reaction of the urine is chiefly due to the presence of the acid phosphate of soda, but although it may conduce to, it is not the chief cause of the acidity. When feebly acid or neutral urine is boiled, more especially if it contains an excess of earthy phosphates, a white cloud, which is frequently mistaken for albumen, falls down. This cloud or precipitate is easily distinguished from albumen by the addition of a few drops of acid, which, while it clears phosphatic, only renders albuminous urine more opaque.

Scherer explains the deposition of the phosphates by boiling on the grounds that heat changes the more soluble neutral into the less soluble basic salts.

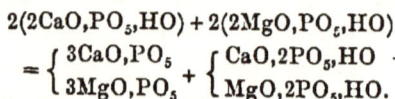

$$2(2CaO,PO_5,HO) + 2(2MgO,PO_5,HO)$$
$$= \begin{cases} 3CaO,PO_5 \\ 3MgO,PO_5 \end{cases} + \begin{cases} CaO,2PO_5,HO \\ MgO,2PO_5,HO. \end{cases}$$

The muddy appearance of alkaline urine is in great part due to the presence of insoluble phosphates, and in the deposit at the bottom, as well as in the scum on the top, crystals of them can easily be detected.

Clear urine containing much phosphate of magnesia, as after a dose of salts, for example, is at once rendered turbid by a drop or two of ammonia which causes a precipitate of the ammoniaco-magnesian triple phosphate to fall down.

Quantitative Analysis.—There are three methods by which the amount of phosphoric acid in urine may be ascertained :—

(a) This is a very common method of analysis, and when followed 50 c.c. of urine are generally employed. The precipitate should be allowed to stand twelve or twenty-four hours before being filtered.

1st. By precipitating the phosphoric acid in the form of earthy phosphates, and weighing the amount, as stated above.

2nd. By a standard solution of the perchloride of iron.

3rd. By a standard solution of uranic oxide.

Of these three methods, the last being now acknowledged to be the best, I shall describe it.

1st. Preparation of standard solution of uranic oxide.

Dissolve 20·3 grammes (314·65 grains) of pure uranic oxide in strong acetic acid, and dilute the solution with distilled water up to 1000 c.c. (32·26 oz.) Each cubic centimetre of this solution will equal 0·005 gramme of phosphoric acid, which Neubauer considers is the most convenient strength.

2nd. Acetate of soda solution.

Take 100 grammes (1550 grains) of acetate of soda with 100 c.c. of pure acetic acid, and dilute with distilled water up to 1000 c.c.

3rd. Prepared paper.

Cut some white filter paper into strips, and saturate them with a concentrated solution of ferrocyanide of potassium. Dry and preserve them in stoppered bottles for use.

Method of Analysis.—Being provided with the above solutions, proceed as follows :—To 50 c.c. of clear urine add 5 c.c. of the acetate of soda solution. Heat the mixture, and while still warm add, drop by drop, the standard solution of uranic oxide (from the apparatus described under the head " Urea Analysis") until a precipitate ceases to form, or, what is still easier, until a piece of the dry, white paper, saturated with a solution of ferrocyanide of potassium, is stained brown, on a drop of the mixture being allowed to fall upon it.

Supposing that 20 c.c. of the standard solution of uranic oxide has been required to precipitate the phosphoric acid in the 50 c.c. of urine, and, as was said, 1 c.c. of the solution is equal to 0·005 gramme of phosphoric acid, it is evident that the 50 c.c. of urine must have contained 0·1 gramme of phosphoric acid. If, then, the patient passed (we will imagine) 1200 c.c. of urine in the twenty-four hours, his daily excretion of phosphoric acid must have been 2·4 grammes.

$$\frac{0\cdot 1 \times 1200}{50} = 2\cdot 4$$

N.B.—Should it happen that the urine contains a precipitate of earthy phosphates, they are to be dissolved by means of a few drops of hydrochloric acid before the standard solution is employed. If the urine likewise contains mucus, filter it after the addition of the hydrochloric acid. If albumen, separate it by boiling.

PHYSIOLOGY.

As already said, phosphoric acid seems to play a most important part in the animal economy, for it not only enters very largely into the composition of the bones and teeth, but is likewise found in the blood, nerves, brain, and muscles. The quantity of phosphoric acid daily excreted with the urine by a healthy adult amounts to about 3·22 grammes (50 grains). Or, according to Parkes, 0·336 grain per lb. of bodily weight.

All the phosphoric acid found in the urine cannot be said to come from the metamorphosis of the osseous, nervous, and muscular tissues, as a large proportion is the direct product of our food. Bread, beef, and potatoes are exceedingly rich in phosphates.

½ lb of beef contains on an average about 30 grains.
"   bread   "   "   32 ,,
"   potatoes   "   "   10 ,,

Fatty, saccharine, and starchy foods contain, on the other hand, comparatively little. The more, therefore, we indulge in bread, beef, and potatoes the more phosphates are eliminated

K

with the urine.(a)   Even the very fæces, after the greater
part of their phosphates have been absorbed during their
passage along the digestive canal, still contain 10, 20, or 40
per cent., the amount varying according to the kind and
quantity of the food and drink.

In the urine of a small but well-made, healthy female child,
aged 17 months, I found on the average of two days 0·832
gramme (12·896 grains) of phosphoric acid in the twenty-four
hours' urine, and calculating the amount for every pound of
the child's weight (21 lbs.), it gives 0·04 gramme (0·62 grain)
per lb., nearly double what the adult passes.   This result was
certainly contrary to what I anticipated, but I have little
doubt it arose from the child being fed on food rich in phosphates (milk, beef, and bread), and its appetite being also
remarkably good.

On another occasion I analysed the urine of an infant
aged six months, which received no food beyond its mother's
milk, and its urine contained the merest traces of phosphoric
acid, not sufficient to be worth a quantitative analysis.   This
child was teething at that time.

Old people's urine in general contains an abundance of
phosphates.   This sample of urine from a man aged 70 gives,
as you perceive, a very copious deposit of earthy phosphates
when it is rendered alkaline with a few drops of ammonia.

The more fluids we drink (cæteris paribus) the more phosphates (of the food) are dissolved and taken into the circulation, and the greater is the amount eliminated with the urine.
Phosphates never entirely disappear from urine, even after
prolonged fasting, so it is perfectly evident that the quantity
passed under such circumstances must represent the amount

(a) After feeding a number of white rats for some weeks on nothing
but bread and milk, their urine became so rich in phosphates that wherever a drop fell it left a white crust of beautifully-formed prismatic
crystals.

furnished by the disintegration of the tissues, more particularly of the nervous and osseous, which are by far the richest in these ingredients.

I must here warn you against falling into the common error of supposing that whenever there is a spontaneous deposit of phosphates in fresh urine, there must necessarily be an excessive elimination, for exactly the contrary may be the case. If, for example, the urine is from any cause alkaline, as was before shown, the earthy phosphates being insoluble in alkaline solutions, they are immediately thrown down. On the other hand, should the urine chance to be abnormally acid, there may be an excessive elimination both of alkaline and earthy phosphates, and yet no precipitate be visible. An excessive elimination has on this account often escaped detection. In fact, there is no possible means of ascertaining the amount of phosphates in the urine except by a quantitative analysis, and this is the more unfortunate, seeing that their quantity, in many cases, furnishes a valuable clue to diagnosis, as well as an important guide to treatment.

As it occasionally happens that a spontaneous deposit of phosphates falls down after a patient has taken a dose of sulphate of magnesia, this circumstance must never be forgotten in cases where we are guiding our treatment by the results of our analyses.

There is still another point to be borne in mind—namely, that many of the drinking waters in different parts of England contain a large quantity of lime, and that persons who have been perfectly free from phosphatic urine in one district may be suddenly attacked with it on going to reside in another where the above-mentioned condition of water exists. This occurs, of course, only in those in whom there is a predisposition to phosphatic affections, either in the shape of gravel or stone.

On examining urinary deposits under the microscope, one occasionally meets with a specimen containing crystals of both uric acid and triple phosphates. At first sight this appears very extraordinary, as the one substance is only found in acid and the other in alkaline solutions. In all these instances, however, the urine is stale, and the simple explanation of finding the two deposits in the same specimen is, that the crystals of uric acid were formed when the urine was fresh and acid, while those of the phosphates appeared later, when decomposition had rendered the secretion alkaline. By boiling such a specimen of urine, the crystals of uric acid disappear in consequence of being transformed into the urate of ammonia; whereas the phosphates remain undissolved.

Phosphates and oxalates are occasionally found associated together.

PATHOLOGY.

Like all other urinary ingredients, the amount of phosphoric acid eliminated with the urine undergoes a marked change in disease; and this, too, irrespective of the influence of a change in diet, which is admitted to be a factor of much importance, seeing that, as before mentioned, certain kinds of food furnish to the system twice, thrice, or even four times as much phosphoric acid as others. The diminution in the elimination of phosphoric acid which takes place in many febrile and inflammatory affections may, no doubt, be in several cases chiefly due to the low diet. In some few, however, it is even much less than can be so accounted for.

Vogel, in a case of pneumonia, found the amount of phosphoric acid reduced during eight days to an average of 1·417 grammes (22·69 grains), less than half the normal amount. On the other hand, it has been ascertained that in inflammatory diseases of the nervous system, notwithstanding the low diet on which the patient may be placed, there is an actual increase instead of a diminution, in the amount of phosphoric acid excreted. This has been noticed to occur both in the idiopathic and traumatic forms of inflammatory action.

Dr. Bence Jones found an excessive increase of phosphates in a case of phrenitis ; while in a case of paralysis following upon injury to the head which was in University College Hospital under the care of my colleague, Mr. Erichsen, I found an equally enormous excess of phosphates. A few days after admission, the patient eliminated no less than 8·749 grammes (135·609 grains) of phosphoric acid in twenty-four hours ; and this, too, at a time when he was taking very little food, and that little poor in phosphates. The urine was at the same period strongly alkaline, and gave on standing a copious spontaneous deposit of earthy phosphates. Gradually, as the patient recovered from the effects of the injury, which was chiefly in the neighbourhood of the fourth ventricle (as diagnosed from the presence of sugar in the urine), the phosphoric acid daily diminished, notwithstanding that his food was more abundant, and of a kind richer in phosphates.

As this case is exceedingly important, I shall here give the result of the analyses, some of which were made by my former pupil, Dr. Pringle :—

Twenty-four Hours' Urine.

Date.	Quantity of Urine.		Reaction.	Deposit.	Phosphoric Acid.		—
	c. c.	ozs.			grms.	grs.	
July 18	952	30·7	Alkaline.	Copious phosphatic.	..	..	..
„ 20	1000	32·2	Do.	Do.	..	..	..
„ 21	2187	70·5	Do.	Copious deposit of mucus and phosphates.	8·749	135·609	..
„ 22	1875	60·5	Do.	Do.	5·020	77·810	..
„ 25	1955	63·0	Do.	Less abundant.	3·357	51·033	Alkaline : 1·119 grms. 17·344 grs. Earthy : 2·238 grms. 34·689 grs.
„ 28	1000	32·2	Do.	Do.	..	..	..
Oct. 15	915	29·5	Acid.	None.	4·57	70·835	..
CONVALESCENT.							
Oct. 20	1375	44·3	Acid.	None.	4·12	63·860	..
Dec. 3	974	31·4	Do.	Do.	2·95	45·725	..

This is an unusually instructive case, both for the reasons already given, and because, while normal urine contains about four times more alkaline than earthy phosphates, their relationship, on the day note was taken of it, was entirely reversed; the amount of the earthy phosphates being double that of the alkaline. The case is still further interesting, from the fact that, after the man had recovered from the effects of the injury, the amount of excreted phosphates actually fell below the normal standard, making it appear as if the body, in order to compensate for the previous loss it had sustained, was replenishing itself with an excess of phosphoric acid from the food. Hassall relates the case of a woman, aged 54, labouring under partial paralysis of the leg, whose urine contained such an amount of triple phosphates that it was thought to be purulent.

This excessive elimination of phosphates in cases of paralysis is exceedingly difficult to remedy—more especially when the paralysis is due to disease in the spinal cord—it being the cause, not the effect we have to combat; for just as albuminuria is but one of the signs of Bright's disease, so is the excess of phosphates in the urine only one of the signs of the severe nerve-lesion ; which must, of course, be treated according to the special requirements of the case. One hint, however, I may venture to give regarding the treatment of these cases, namely, that wherever there is an absence of inflammatory symptoms, and the paralysis appears due to a degeneration or mal-nutrition of the nerve tissue, one of the most successful lines of treatment is to restore to the system the phosphoric acid it has lost. In such cases, phosphoric acid often proves of even more signal service when combined with small doses of strychnine, which, as is well known, is one of the best nervine tonics in the Pharmacopœia.

Where phosphoric acid is taken internally, either in a free or in a combined state with soda, it reappears in the urine in the form of phosphate of potash (Böcker). So that, if at any time we wish to diminish the amount of potash and increase the quantity of soda in the system, all that is required is to give phosphate of soda.

In mania, Dr. Sutherland concludes from the results of several analyses that were made for him by Dr. Beale, that the phosphates in the urine are increased during the acute paroxysm, and diminished during the subsequent state of exhaustion. He thinks, also, that there is a diminution of the phosphates in the urine in acute dementia, as well as in the third stage of paralysis of the insane.

In cases of delirium tremens, Dr. Bence Jones found the phosphoric acid diminished in the urine.

On some rare occasions the urine has been observed to be phosphorescent at the moment of emission, the falling drops being luminous in the dark like those of phosphorescent sea water. The cause of this appearance is as yet unknown. These cases generally occur in winter.

In diseases of the bladder the urine is often observed, at the moment of its emission, to be loaded with crystals of the ammoniaco-magnesian phosphates; a circumstance which has not unnaturally given rise to the false impression that in all such cases the phosphates are in excess. Whereas it occasionally happens that in some of these very cases, instead of the elimination of the phosphates from the system being excessive, it is actually below the normal standard. The explanation of this fact has already been given when speaking of the physiological relations of phosphoric acid. It will be remembered it was then pointed out that the earthy phosphates are always precipitated in alkaline liquids, whether they chance to be in small or large quantity. Hence the appearance of a deposit of phosphates in alkaline urine freshly emitted from a diseased bladder is no criterion of their absolute quantity. Be this, however, as it may, the great point of practical importance

remains the same—namely, that all such cases merit close
attention, and that the best energies of the physician must be
directed to preventing a deposition of the phosphates taking
place during the urine's sojourn in the bladder. This point
we shall have occasion to recur to when treating of the utility
of benzoic acid and opium in the alkaline urine of vesical
disease.

In Bright's disease and the various other forms of albuminuria, the phosphates, like most of the urinary salts,
undergo a considerable diminution on account of the disease
of the kidneys preventing their elimination.

In gout they are also said to be diminished.

In cases of rickets the phosphates in the urine, especially
the phosphate of lime, are greatly increased (Lehmann,
Neubauer, and myself)—a fact which may be turned to most
useful account in practice.

Some physicians have erroneously imagined that as there
is an excess of phosphates in the urine in cases of rickets,
there can be no lack of them in the system ; whereas it is
exactly the reverse, the excess in the urine being obtained
from the bones, which gradually become softer and softer as
the phosphate of lime is drained from them. In order to
compensate for this loss, therefore, we ought to give to the
child bone earth (superphosphate of lime). The dose of this
substance for a child of three or four years old is five grains
three times a-day. As the salt is very insoluble, I, like many
others, have had it mixed with various substances, milk, cod-
liver oil, glycerine, mucilage, etc., without, however, succeeding in obtaining a good mixture. Indeed, the only true solutions
that I have obtained were made by dissolving bone-earth in
phosphoric acid, the solution being afterwards rendered clear
by the addition of a drop or two of sulphuric acid. They
contained five grains of phosphate of lime to the drachm of
syrup. These " syrups of phosphate of lime " are at first

colourless, as well as clear, but gradually after some weeks they become of a rich sherry wine colour in consequence of the gradual action of the acid upon the syrup. There is one great objection to their use, however—namely, that they are so acid that many children refuse to take them.

It is better, therefore, in the majority of cases to content oneself with giving the finely powdered bone-earth, either in a mixture along with cod-liver oil, or in milk, without any attempt at artificial solution, trusting entirely to the dissolving powers of the digestive secretions for its solution and absorption.

This plan of treatment cannot be too widely known, for it is no exaggeration to say that of the numerous cases of rickets I have treated with the superphosphate of lime, I do not at the present moment recollect a single instance in which it did harm; while, on the other hand, I remember many that were benefited, and several that might be said to have been actually cured, by it. The word cured is here used advisedly, for I have had children of from two to three years of age brought to me at the Dispensary and at the Hospital who could neither stand nor walk, although they had already been under the usual routine of treatment, and these same children after six weeks' administration of the superphosphate of lime have walked into the waiting-room, with no other assistance than that of their parent's hand.

It would be wrong, perhaps, to attribute the entire benefit to the superphosphate of lime; but there can be little doubt that the other means were only auxiliary, and that to the superphosphate is chiefly due the successful results.

The circumstance which had the greatest effect in inducing me to give in this way bone-earth was my having noticed that the peasant women in the villages round Giessen give their children lime along with their milk. On asking why they did so, the answer invariably was, "to make their legs straight"—a point of no mean importance in that country,

where it is not the length, but the breadth of the female
garments which constitute their chief beauty, neither petti-
coats nor dress being under any circumstances permitted to
reach below the knee.   Whether or no the handsomely-
formed limbs of the fair sex in the valley of the Lahn and
its neighbourhood ought to be attributed to the judicious
maternal administration of lime may be a matter of opinion ;
but nevertheless, be that as it may, the two facts here related
remain undeniable.

To children of the upper class there is fortunately another
mode of supplying their bones, muscles, and nerves with
phosphates—namely, by giving to them bread raised with
"Brotopian" instead of cream of tartar.   This substitute (an
American preparation) makes bread exactly like that made
with cream of tartar, and it has the additional advantage of
containing nothing but what is to be found in beef-steak, corn,
wheat, and other cereals.   It is also said to have the still
further advantage of making the cakes, biscuits, or bread
more palatable, so that children eat them without hesitation.

Mollities ossium is another of those diseases in which the
phosphates are greatly diminished in the bones, while at
the same time they are increased in the urine (Beale);
and there can be but little doubt that the same treatment which has been said to be so useful in rickets
will be equally serviceable here.   Indeed, wherever there
is an excessive elimination of phosphates from the system, without any deposition of them taking place in the
urinary passages, the best line of treatment to adopt is to *give
more phosphates ;* and that, too, for the same reason that if
you had to bring to a person a pint of water in a pail with a
hole in its bottom, which you had not the means of stopping
up, you would put much more than a pint of water into the
pail in order to allow for the loss likely to be sustained in
the transit.

When the excessive elimination of phosphates, on the other hand, is attended by a deposition of a part of them in the urinary passages, in the form of sand, gravel, or stone, an entirely different line of treatment ought to be adopted. .

Phosphatic Gravel and Calculi.

Earthy phosphates are frequently deposited from urine before its emission from the bladder. The deposition may take place either in the kidney, and give rise to a renal calculus; or, what is still more common, but fortunately less dangerous, it may take place in the bladder, and produce an ordinary stone.

The pathology of renal and vesical phosphatic calculi is nevertheless very different. Thus, renal phosphatic calculi arise in one of two ways—either from the urine containing an excess of phosphates, in consequence of a vice in the tissues which contain them (bones, brain, &c.); or from the urine being alkaline at the moment of its secretion in consequence of being secreted from abnormally alkaline blood.

In the case of vesical calculi, on the other hand, in addition to the causes just mentioned, there exists still another fruitful source of origin in the numerous diseased conditions of the bladder itself. Thus, for example, whatever induces retention of the urine and favours the decomposition of its urea, as occurs when a patient is labouring under paralysis of the bladder, or stricture may give rise to a vesical phosphatic stone.

Diseases of the digestive organs, which are among the most common causes of the other kinds of gravel and stone, also exert a particular influence over the formation of phosphatic calculi. And the disease affecting these organs has not unfrequently so masked the symptoms produced by the stone that its presence has remained undetected until revealed by the post-mortem examination.

A specimen of renal calculus, sent to me by Dr. Roome

while he was Physician to the Dartmoor Prison, furnishes a
most striking illustration of the correctness of this remark.
The following is the history:—The patient, a convict, aged 60,
had lived a very dissipated life prior to his conviction, and
during the two years succeeding his conviction he suffered
from gastric derangement, without presenting any symptoms
which would have led to the suspicion of renal disease. This
is the more surprising, seeing that no less than five ounces and
a half of phosphatic calculi were removed after death from the
right kidney (the left contained none), and one of the calculi
alone weighed three ounces and three-quarters. The man
died from cancer of the pylorus, which induced a stricture of
the orifice hardly admitting of the passage of a crow-quill. If
any renal symptoms existed in this case, they were entirely
masked by the more serious disease of the stomach. All the
calculi consisted of triple phosphates.

As regards the first step in the formation of calculi, it may
be laid down as a general law that *their formation in any
particular part of the urinary system is due to some special local
cause.* If a calculus begins in a urine tube, it is on account
of the tube getting blocked up with mucus, epithelium, or a
little blood. Or it may also arise from the tube being temporarily constricted in some parts of its course in consequence
of inflammatory action. Once, however, that the deposition
of urinary sediment has begun, the original cause is soon lost
sight of, and the calculus goes on forming round its nucleus
quite independent of the local condition which first called it
into existence.

When a phosphatic calculus, instead of beginning in a urine
tube, originates in the pelvis of one of the kidneys, there may
be no absolute obstruction to the flow of the urine, but merely a
slight interruption, as from a fragment of mucus adhering to the
side of the lining membrane of the pelvis. For, when a predisposition to the formation of a calculus exists, it requires but a

very trifling cause to call it into operation. Almost anything may form a nest round which the stone will gradually become deposited.

This is no fanciful description of the formation of renal calculi, as you may see by taking the trouble to examine the series of examples now on the table, each of which bears evidence of the truth of the statement. In fact, the assertion might be proved without any specimens at all, from the simple circumstance of the frequency with which we meet with renal calculi limited to one kidney in patients predisposed to stone. That fact of itself clearly shows that the deposition of the stone in any one particular spot must be entirely due to some local cause. Thus, for example, we have here a kidney, which contains no less than fifty-nine uric acid calculi; and yet neither in the bladder nor the opposite kidney of this gentleman was a single stone to be found. The deposition of all these calculi in one kidney must therefore have been entirely due to some local cause. At first sight, the cause does not appear very evident, nor is it apparently easy to account for the fact that although at least fifty of these calculi are not bigger than barleycorns, none of them have found their way into the ureter, and from thence onwards into the bladder. The reason of the last fact is, however, apparent when the specimen is carefully examined, and as it will be easier to explain the former proposition after we have discussed the latter fact, we shall consider the last first. The largest stone, which, from its position and size, is evidently the one which was first deposited, is of a very peculiar conical shape, and the apex of the cone projects into the orifice of the ureter, and forms a movable plug, which is, in fact, a self-acting valve. No sooner does the pelvis of the kidney become filled with fluid, than its distended walls cease to embrace the stone, and consequently the fluid flows past the valve until the pelvic walls again come into contact with it. The separation of the walls from the valve, although being sufficient to allow liquid to pass, does not appear to be sufficient to admit of the escape of even the smallest of the stones. Now exactly the same thing as we have done artificially with water must have been daily done with the urine.

Gradually as the urine was secreted, it would distend the pelvis of the kidney and separate its walls from the valve-shaped calculus ; but no sooner would this occur than the accumulated urine would make its escape. So that every now and again there would be little accumulations of urine, and during the period of these little accumulations not only would the smaller calculi form, but the larger one, the plug, would gradually increase in size. In this case, it is perfectly evident from the immense number of the calculi that the predisposition to stone existed; and from their being all limited to one kidney it is just as evident that it was some local cause which induced them to select the left instead of the right kidney. In Dr. Roome's case, again, it was precisely the reverse : all the calculi were found in the right kidney ; not one existed in the left ; which is additional evidence in favour of our views.

In my own case, moreover, it is further shown, by the absence of all calculi both from the right kidney and from the bladder, that, although the predisposition to stone existed, yet without some local cause no stone would have been deposited in the urinary passages of the patient—a most suggestive fact, as showing the importance of paying attention to the proper performance of the renal function. Who can tell what a trifle at first induced the deposition of the uric acid nucleus, which ultimately became the plug and caused the formation of all the rest ?

The foregoing general principles are not alone applicable to the pathology of phosphatic, but to all varieties of renal calculi—oxalic, uric, xanthic, etc.

As regards the first step in the formation of vesical calculi, it may, with equal justice, be said that whenever the predisposition to stone exists, it requires but a very trivial cause to bring it into operation ; for, as with renal calculi, almost anything may act the part of a nucleus round which the stone will become deposited. An insignificant particle of mucus, a microscopic portion of albumen, or an epithelial cell, is all that is requisite for the purpose. In the vast majority of vesical calculi, indeed, it is impossible to detect the primary point round which the urinary salts became adherent ; but every now and then we come upon calculi with well-marked nuclei, some of which are not only quite foreign to the urinary passages, but even foreign to the body itself. Thus, for example, this large ammoniaco-magnesian stone (No. 3425 of Liston's collection in the museum of University College), which was extracted from the bladder of a man aged 60, has a piece of beeswax for its centre ; while three other small calculi (No. 3478), removed from the bladder of a man aged 70, have barleycorns as their nuclei. It is perfectly clear these substances were not generated spontaneously in the urinary passages, and it is highly probable also that they did not find their way there quite accidentally ; but with the mode of entrance we have at present nothing to do ; so we may turn our attention to this other vesical calculus (No. 3515 of the same collection), which was removed from a woman's bladder. The nucleus of this stone consists of a piece of bone —a fœtal tibia, in fact—which found its way by ulceration into the bladder from the abdomen, the case having been 'one of extra-uterine fœtation. Any solid substance, indeed, may prove the starting-point of a vesical stone, just as a thread, piece of string, or anything else, may form the nucleus of a crystal of sugar-candy.

In connection with the origin of calculi, there is another point particularly deserving of attention—namely, that local causes are the most potent in the production of stone, and that even where the predisposition to it exists, without some local cause no stone is likely to form.

Vesical phosphatic calculi are frequently induced by inflammation of the bladder. The inflammation gives rise to a copious secretion of mucus, which, acting as a ferment, causes the urea to decompose, while the urine still sojourns in the bladder. The decomposition of the urea leads to the evolution of ammonia, which, in its turn, deposits the earthy phosphates, and the gradual formation of a stone is the result.

Many phosphatic calculi have a uric acid nucleus, while some few have oxalate of lime for a centre. The deposition of phosphates, and consequent increase of the stone in such cases, is due to one of the four following causes :—

1st. Although the patient has recovered, either by time or by treatment, from the uric or oxalic acid diathesis,(a) the stone that has already formed still remains in the bladder, and immediately, or at some future date, causes such an amount of irritation as to induce an excessive secretion of mucus; which mucus, acting as a ferment, brings about the decomposition of the urea, and consequent formation of the alkaline carbonate of ammonia; which, in its turn, induces, as in the preceding examples, a deposition of the earthy phosphates normally present in the urine.

In this case the deposition of the phosphatic calculus is not due to a vice in the general system, but the product of an entirely local cause.

2nd. A patient with a small uric or oxalic acid stone in the bladder, which induces no irritation whatever, is on some occasion or other seized with retention of urine. It may be from stricture, enlarged prostate, paralysis of the bladder, or

---

(a) The word diathesis is here employed merely to save a multiplicity of words, not from any special significance being attached to its meaning, which is now pretty well understood.

any other cause liable to induce it. In consequence of this retention the urine becomes decomposed, and a deposition of phosphates takes place round the nucleus, the presence of which until now was of no importance whatever. The deposition of phosphates is, however, copious, and the stone rapidly increases in size, and becomes of itself a source of irritation. Here, then, neither the original nucleus nor the constitution of the patient is to blame, but a local cause quite independent of either.

3rd. A deposition of phosphates may take place round a nucleus of uric acid, oxalic acid, or anything else, not on account of its setting up irritation, nor because of the urine being retained and decomposed, but from there being an excessive elimination of phosphates by the kidneys. In common language, on account of the patient labouring under the phosphatic diathesis.

Lastly. A phosphatic calculus may form round a uric acid or other nucleus without any of the already-mentioned conditions existing, but simply and solely from there being an absence of that normal acidity of the urine which is essential to the retention of the earthy phosphates in solution until its emission from the bladder.

The diagnosis of cases of phosphatic calculi, renal as well as vesicle, is not only greatly aided by an examination of the urine, which may be said to contain, as a general rule, less phosphoric acid than normal; but even the differential diagnosis of the various causes inducing the calculi is also aided by an examination of that liquid.

1st. In cases arising from an excessive elimination of phosphates the reaction of the urine is normal, and it has no ammoniacal odour, nor does it deposit an excess of mucus.

2ndly. In those arising from an abnormal alkalinity of the blood the urine, although alkaline at the moment of its emission, either immediately after meals or during the whole

L

day, has no ammoniacal odour, nor does it deposit mucus in excess.

3rdly. In the first and second class of cases the urine is ammoniacal as well as alkaline when passed, and usually yields a copious deposit of mucus.

Lastly. The first and second class of cases are readily distinguished from each other by their histories.

Besides the triple or ammoniaco-magnesian phosphate, which is the most common kind of phosphatic calculus, there is also another variety composed solely of the simple phosphate of lime. This is, comparatively speaking, a rare form of stone; and fortunately, its appearance is so very peculiar that it can be readily recognised without an analysis. Its external surface is smooth, polished, and shining like glass. It is exceedingly brittle, and has the peculiar tendency to break into regular layers like an onion.

Treatment.—From what has been already said, it must have been seen that the various classes and kinds of phosphatic calculi each demand a separate mode of treatment, and that the old method of treating all kinds of phosphatic caculi alike is radically wrong. The pathology is the guide to the treatment.

In the first class the stone itself has to be attacked; in the second class the cause of the retention is to be removed; in the third class the vice in the tissues has to be got rid of; while in the fourth the normal acidity of the urine is to be restored. One and all of these are more or less within our power.

First, as regards the class of cases where the deposition of phosphates depends upon a decomposition of the urine taking place in the bladder. That we have seen may originate in two ways; and, as we shall again have occasion in another lecture to refer to the mode of treating both of the causes above specified, we shall for the present take it for granted that the cause has been removed, and that all that has now to be done is to get quit of the gravel or stone that has already formed in the bladder.

The common way to remove a vesical stone is, of course, to cut or crush it; but many persons, and especially old people, in whom the phosphatic calculus is most common, object to either of these operations being performed. What, then, is to be done? Fortunately, in the case of phosphatic calculi, the physician has much in his power, for of all calculi the phosphatic are the most soluble; and if only gravel has as yet formed the chance of getting it removed by dissolution is very great. The simplest mode of treating such cases is to keep the urine constantly acid by the administration of the nitro-muriatic acid three times a day, after food, and to irrigate the bladder by means of the double irrigating catheter. This is a much easier operation than most persons who have not tried it imagine. The patient being placed on his back, the double catheter is passed into the bladder, the urine allowed to flow out, and then a caoutchouc tube, at least four or five feet long, is slid over the end of the catheter, while the free end of the tube is dipped into a can containing water acidulated with nitro-muriatic acid. The can having been previously placed at a convenient height above the patient—on the top of the bed-post, for example—the caoutchouc tube is to be so arranged that, when the patient is ready, the acidulated water will flow in a regular stream down through the tube, along the catheter, into the bladder, and from the bladder back again through the other division of the catheter, along another caoutchouc tube attached to its opening, into a pail placed beneath to receive it. If the acidulated water be kept at a temperature of about 99° F., the stream through the bladder may be kept up for a long time without causing the slightest discomfort to the patient after the first sitting or two.

When the top pail gets empty it is refilled from the other as often as is thought necessary. This treatment being continued for a certain time every day, the gravel is gradually dissolved without causing the patient any suffering.

In the third set of cases strict attention should be paid to the general health, and an effort made to improve it, and thereby restrain the elimination of phosphates from the system by a judicious selection of tonics. If this is not effectual and the daily waste excessive, phosphoric acid is to be administered, for the reasons previously given.

In the fourth class of cases, where the vice lies in the abnormal alkalinity of the blood, the strong mineral acids ought to be administered. Vegetable acids, on the other hand, are to be avoided, from the fact of their being transformed into carbonic acid in the system, which only tends to increase the mischief. In general, it is sufficient to give the acid after food ; but in very bad cases it is sometimes necessary to administer acid tonics also between the meals. In many of these cases the patients perspire freely, and the sweat is excessively acid, which in some measure accounts for the diminution of acid in the blood. To this class of patients I find much benefit arises from the use of nitro-muriatic acid in conjunction with gentian after meals, and a dose of sulphuric acid, to diminish the perspiration, between them. The patient's diet must also be carefully regulated : the animal food increased, the vegetable food diminished. The reason of this is plain enough, if what was said in the first lecture regarding the influence of diet on the reaction of the urine be considered. Herbivorous animals have alkaline, carnivorous animals acid urine; so that the more the patient resembles the carnivorous animal in his diet the less alkaline his blood, and the more acid his urine will become.

In trying thus to check the formation of phosphatic gravel or stone, care ought at the same time to be taken not to run to the opposite extreme ; and, by increasing the acidity of the system too much, produce a deposition of gravel or stone of another kind—uric acid, for example. The safeguard against falling into this error is found in the condition of the urine. As soon as the normal acidity of the urine becomes permanently restored, the acid treatment and animal diet ought to be stopped.

The following case is an admirable illustration of the benefit to be derived both from the means of diagnosis and the line of treatment I am now striving to inculcate. This specimen of urine is from Charles R., a patient at University College Hospital, now under my treatment for renal phosphatic calculus in the left kidney. The history of the case is briefly as follows:—Charles R., whose age is 49, came to the Hospital on December 1, 1863, complaining of pain in the region of the left kidney, which was increased by pressure applied both anteriorly and posteriorly. He said he had suffered from it during the last fifteen years. But what troubled him even more than the pain was the constant desire to make water. Sometimes he had to make it as often as six or seven times within a quarter of-an hour. As a rule, however, he micturated only once in every two hours; but after each micturition had to go three or four times before the desire entirely ceased. At last only a few drops passed. This was attended with great discomfort. During the fifteen years he had been so afflicted, he had passed on one or two occasions a chalky-looking substance; and a year and a-half ago a little blood came away with the urine. As some of these symptoms gave rise to the belief that there might be a vesical calculus, I sent him in to my colleague, Sir Henry Thompson, and had the bladder sounded; but no stone could be detected. On analysis the urine was found to be alkaline, and deficient in phosphates; this fact, in conjunction with the history and symptoms, led me to the diagnosis of phosphatic renal calculus, arising from an abnormal alkalinity of the blood. Such being the diagnosis, the patient was ordered to take the following mixture three times a day :—℞ Acidi hydrochlorici

dil., ʒss.; acidi nitrici dil., ₥v.; infusi chiratœ, ʒss. M.
Already, on December 4, the patient made urine less
frequently, and stated that he felt better. The quantity was
1350 c.c. (43¼ ozs.); the specific gravity 1019; the amount of
phosphoric acid 4·725 grammes (73·28 grains). The same
treatment was ordered to be continued : and when I saw
the patient on the 14th (ten days later) he spontaneously
declared that he had never felt so well during the last fifteen
years; in fact, he stated that during that whole time
his life had been a burden to him. The reaction of the
urine was now acid; and, although its quantity had been
reduced to 1168 c.c. (38 ozs.) in the twenty-four hours, the
phosphoric acid had increased to 5·46 grammes (84·63 grains),
the increase being of course caused by its being no longer
deposited in the urinary system. Being desirous of seeing if
this improvement was entirely due to the medicine, I proposed
discontinuing it; but at the urgent request of the patient had
to content myself with merely diminishing the dose, as he declared it was the only remedy that had ever done him good,
and he feared to fall back into his former miserable state. On
December 21 he returned, saying he was not quite so well, and
requested to be allowed more of the medicine. The urine
was again analyzed by my class-assistant, Mr. Mason, in order
to ascertain if the patient's statement was correct, or if he
only imagined he was worse. The following is the result of
the analysis :—Quantity, 1240 c.c. (40 ozs.); specific gravity,
1022. Reaction slightly alkaline. The amount of phosphoric acid, 4·65 grammes (72·07 grains). The urine gave a
considerable deposit; which, when examined microscopically,
was found to contain prismatic crystals of triple phosphate,
and a few octohedral crystals of the oxalate of lime. The
examination of the urine, therefore, quite confirmed the
patient's statement regarding his not being so well as on the
full dose of the remedy.

LECTURE VIII.

OXALIC ACID DIATHESIS—OXALURIA—ITS CLINICAL SIG-
NIFICANCE AND TREATMENT—MULBERRY CALCULI.
SULPHATES; THEIR CHEMISTRY, PHYSIOLOGY, AND PATH-
OLOGY.

CHEMISTRY.

THE well-known poison, oxalic acid, which so closely resembles
the sulphate of magnesia in appearance as to be frequently
mistaken for it, although one of the most powerful of the
organic acids, appears to be an almost constant component of
the blood.   It is, however, rarely present in sufficient quantity
to give rise to any inconvenience ; and is only excreted from
the kidneys in appreciable amount under certain abnormal
conditions of the system, or after having been taken in medicinal or poisonous doses.   Its presence in solution is easily
recognised by the property it has of giving with any soluble
salt of lime a white precipitate possessing well-defined crystalline characters.   As usually met with in the urine, the
oxalic acid is already in combination with lime, and fortunately
it is so ; for the quantity is generally so small that, were it not
for the facility with which the crystals of oxalate of lime are
recognisable with the microscope, its presence, in the majority
of instances, would escape detection.

The oxalate of lime is most readily detected in the following
manner :—Put about six ounces of freshly-passed urine into
a conical glass ; and, after it has stood for eight or ten hours,

place a drop of the sediment under the microscope; and, as the crystals are exceedingly minute objects, examine with a high power, at least a quarter of an inch. If the urine has deposited some mucus, the mucus ought also to be examined; for the oxalate of lime crystals are often entangled amongst it, and are thereby prevented falling to the bottom of the vessel. The various forms assumed by the crystals, as seen under the microscope, are represented in Figure 14.

FIG. 14.

Oxalate of lime from human urine.

They may be divided into four well-marked groups—the octohedron, the dumb-bell, the irregular disc, and the well-defined diamond-shaped crystal. Of all these four forms, however, it is the octohedron alone which is characteristic of the oxalate of lime. Lithates may be found as dumb-bells, carbonate of lime as discs, and uric acid as small diamond-shaped crystals, whereas the only other urinary ingredient which ever crystallises in octohedrons is the chloride of sodium; but there is no possibility of error on that score, as it is so extremely soluble as never to appear as a spontaneous deposit in the renal secretion. Furthermore, it may be remarked that, even in those cases where the oxalate of lime assumes the shape of dumb-bells, discs, or diamonds, there usually happens to be present at the same time several octo-hedrons ; so that the danger of falling into error regarding the nature of the deposit is very slight.

The only two sediments that I can imagine it possible the oxalate of lime can be confounded with are those of triple phosphates, and uric acid. The prisms of the former may, perhaps sometimes to beginners, appear almost octohedral in shape ; and the diamonds of the latter look so small as to lead to the idea of their being oxalates. But even then a trifling chemical manipulation will set the point at rest; for by the addition of acetic acid the phosphates will be dissolved, and by the addition of caustic potash the uric acid will disappear, while the oxalate of lime crystals, on the other hand, resist the action of either of these agents.

When the presence of oxalic acid in the urine in a free state is suspected, to a portion of the liquid is to be added a few drops of acetic acid, and then a solution of a salt of lime— chloride, for example—and the mixture placed aside for a few hours to crystallise. The sediment is then to be examined microscopically as above described.

Quantitative Analysis.—To make a quantitative estimation of the oxalic acid in human urine is extremely difficult; and, for clinical purposes, one usually contents himself by guessing at the amount from the number of oxalate of lime crystals present in the field of the microscope. But in cases where it is deemed advisable to make a closer approximation of the amount of oxalic acid daily excreted by the kidneys, the fol-lowing method is to be adopted. The urine for the twenty-four hours is collected and carefully measured. To a sample equal to 200 c.c. is then added some milk of lime, and the mixture evaporated to dryness. The residue, after being acidulated with acetic acid, is next extracted with repeated portions of alcohol, and the alcoholic solution shaken up with about 50 per cent. of ether, which throws down the oxalate of lime as an insoluble deposit. This is collected, washed, dried, and weighed, and the whole amount passed in the twenty-four hours estimated therefrom in the usual way. (See the formula under the head of uric acid.)

PHYSIOLOGY.

It is still a debated question whether or not oxalic acid is a normal constituent of the human frame. Böcker says he passes ·092 gramme (1·42 grains) in his twenty-four hours' urine ; but, whether this amount is furnished by the disintegration of the tissues, or is the direct product of the food, it is extremely difficult to say. Oxalic acid is widely distributed throughout the vegetable kingdom. Many fruits, and the majority of vegetables, contain considerable quantities of it ; whilst in rhubarb, and sorrel it exists in great abundance. Moreover, many foods which do not naturally contain oxalic acid are readily transformed into it in the system. Among these may be mentioned starch, sugar, vegetable acids, salts of the vegetable acids, and carbonated alkalies. It is, therefore, almost impossible to prevent its finding access to the body ; and, consequently, my own impression is that it ought to be regarded as one of the normal ingredients. While making this assertion, however, I wish specially to guard against it being supposed that I am one of those who look upon the presence of oxalate of lime in the urine as a matter of trivial importance ; for, on the contrary, as *normal urine never contains any sediment*, the appearance of oxalate of lime or of any other crystals within twelve, or even twenty-four hours after the urine has been passed, is a sure indication of the existence of an unhealthy condition of the system—temporary it may be, but nevertheless abnormal.

Oxalic acid to be normally eliminated from the body ought, like all the other natural ingredients of the frame, to be excreted in an invisible form ; and I strongly suspect that it will

yet be found that the healthy human body does in reality
daily eliminate a small quantity of oxalic acid in a soluble
and consequently invisible form; for every now and then, on
evaporating what has appeared to be normal urine to nearly
dryness, I have met with crystals of the oxalate of urea
(Fig. 15), both the oxalic acid and the urea having remained

FIG. 15.

Oxalate of urea from human urine.

invisible until the liquid was highly concentrated.   Oxalic
acid has in some cases been detected in human blood. Garrod,
for example, in searching for uric acid in the serum, has occasionally encountered crystals of the oxalate of lime.

When oxalates are given in large doses medicinally, as well
as when a poisonous quantity is taken, either suicidally or by
accident, the blood becomes so saturated with the acid that
leeches fall off and die ere they have been many minutes
attached to the skin, and before they have sucked beyond a
very small quantity of blood. I have myself observed this on
two separate occasions.

Oxalic acid appears to be manufactured even in the human
body; for, whenever the metamorphosis of some of the animal
products is interrupted, a deposit of oxalates makes its appearance in the urine.

It is now generally admitted as a recognised fact, that urea is the last product of the retrograde metamorphosis of the nitrogenised tissues; and also, that it is in the form of urea that the greater part of the albuminoid group of foods is excreted from the system. Uric acid, on the other hand, is supposed to be one of the intermediate products of this metamorphosis. For three reasons :—

1st. Because, by the action of permanganate of potash on albuminous substances, uric acid may be artificially obtained;

2ndly. Because, by the further action of permanganate of potash, uric acid is transformed into urea ; and

3rdly. When uric acid is taken into the system, it reappears in the urine in the form of urea ; whereas, when urea itself is taken, it passes through the body unchanged.

We may now add that oxalic acid is also one of the intermediate products in the retrograde metamorphosis both of the tissues, and of the albuminoid group of foods ; for, as is well known, although uric acid, when perfectly oxidised, is completely changed into carbonic acid and urea, when anything intervenes to interrupt the oxidation process, the products are carbonic acid, urea, and oxalic acid. The same transformation also occurs in the body; for, whenever an excess of uric acid is taken internally, or injected into the blood, it appears in the urine partly in the form of urea and partly as oxalic acid. By perfect oxidation, on the other hand, both in and out of the body, uric acid is completely transformed into urea and carbonic acid. Hence we have no difficulty in comprehending why a deposit of oxalates is so frequent a concomitant of diseases of the respiratory organs.

There are several other animal substances besides uric acid that are convertible into oxalic acid. Thus, creatin, when acted on by the permanganate of potash in alkaline solutions, yields oxalic acid. Leucin, tyrosin, and guanin also yield oxalic acid, and so does lactic acid. The neutral fats, too, by lengthened heating with nitric acid, are, among other things, transformed into oxalic acid.

It would therefore seem that, while the amount of oxalic acid present in the healthy animal body may be said to be derived directly from many vegetables and fruits, and indirectly from the whole three groups of food—albuminoid, fatty, and saccharine,—that met with in disease is not only the combined derivatives of all the different kinds of food, but also the direct product of the interrupted retrograde metamorphosis of the various animal substances above alluded to.

PATHOLOGY.

The term oxalic acid diathesis, or, as it is now usually called, oxaluria, is given to a class of cases in which the urine during days, weeks, or even months contains a deposit of oxalate of lime. This deposit is usually accompanied by a peculiar but well-marked train of symptoms, which have recently received special attention from Begbie, Beneke, Frick, and others. The symptoms are mostly of a nervous and dyspeptic character, and although not dangerous in themselves, are nevertheless deserving of the closest attention of the physician ; for when left unattended to they oftentimes drift into those of confirmed hypochrondriasis, and end by rendering the unfortunate patient little else than a nuisance, both to himself and his friends. Supposing even that this form of the affection is escaped from, another still more dangerous, though less distressing, is sooner or later sure to arrive—namely, the formation of gravel or stone.

Oxaluria is more frequently met with in men than in women, but the latter are not entirely exempt from it. Moreover, it is most common among persons in the prime of life, and in the better classes of society. Its most prominent symptoms are disordered digestion, irregular and capricious appetite, flatulence, and a feeling of oppression in the chest after eating ;

irritability of the heart, inducing palpitation on the slightest
exertion or excitement. After a time the disposition becomes
morose; and the merest trifles are sufficient to disconcert the
patient. Both resolution and activity are wanting; and the
daily routine of ordinary business becomes irksome in consequence of every trifling mishap being magnified into a serious
evil.

At this stage the disease is still amenable to treatment; and
by proper attention to diet, fresh air, exercise, and a regular
course of acid tonics, the oxaluria may be completely removed
in a comparatively short space of time. If, however, this
stage of the affection is allowed to pass unheeded, the patient
soon begins to suffer from heat or pain in the kidneys,
shooting or extending deep down in the loins, and irritability
of the bladder, inducing frequent micturition. At the same
time, the urine has a tendency to throw down a somewhat
copious deposit of white mucus, entangled among which
may be found numerous crystals of the oxalate of lime. In
the most inveterate cases the patients complain of being impo-
tent, and suffer from the usual train of symptoms associated
with stone in the kidney or bladder.

The presence of oxalates in the urine denotes, of course, the
pre-existence of oxalic acid in the blood; and it is now generally conceded that the dyspeptic, as well as the nervous
symptoms just alluded to, are the direct result of the chronic
poisoning of the brain and spinal cord by the minute quantities of oxalic acid in the circulation. The renal and vesical
symptoms, on the other hand, are no doubt due to the local
mechanical effects of the oxalic deposit in the kidneys and
urinary passages.

Beneke thinks that there is still another class of evils due to
oxalic acid blood-poisoning. He believes that the oxalic acid
circulating in the blood dissolves out from the tissues, and
removes from the system the phosphate of lime, thereby

diminishing the activity of the various organic processes of which life is the manifestation. Whether or not this be in reality the case, I am not prepared to say ; all I know is, that an abnormal deposition of phosphates, in the urine is not an unfrequent accompaniment of oxaluria. In these cases the urine is either alkaline, neutral, or, at the very most, feebly acid at the moment of being passed. On boiling, it deposits a quantity of amorphous phosphates, which are again re-dissolved on the addition of acetic acid. In such cases of mixed phosphatic and oxalic deposit, a ready way of ascertaining how much is due to the presence of phosphates, and how much consists of oxalates, is to add acetic acid to the sediment. The acetic acid, by dissolving away all the phosphates, allows the amount of the insoluble oxalates to be estimated.

Again, in another class of cases which is much more common, the deposition of phosphates apparently succeeds the oxaluria, for one constantly encounters mulberry (oxalic acid) calculi with an external coating of phosphates upon them. As this, however, may, in many instances, be due (as was pointed out in the lecture on phosphates) to a mere local cause, I beg to direct your attention to this phosphatic stone (No. 3462 of Liston's Collection in University College Museum), which beautifully illustrates the fact that it is not always so, and that a phosphatic may both precede as well as succeed an oxalic acid diathesis. The surface of the section is perfectly snow-white, except at a point nearly equidistant from the centre and the circumference, where a dark circle, one-twelfth of an inch in thickness, runs through the stone. This line of inky blackness, forming such a striking contrast to the snow-white phosphates, is entirely composed of oxalate of lime. Thus it is seen that, even in a urinary concretion, the clinical history of a patient may be read as truthfully as the geologist can trace the early history of the earth in the different strata forming its crust.

This patient's general constitution of body was phosphatic; and for months—aye, probably even for years—he suffered from a deposition of phosphates in the bladder. But suddenly, and for a time, either through treatment or otherwise, a change took place in his system, and the phosphatic deposition was arrested. During this period, however, the patient laboured under another evil. His blood was loaded with oxalic acid, which his kidneys daily and hourly eliminated in the form of oxalate of lime, which gradually deposited itself round the phosphatic stone. By-and-by, however, he got rid of his oxaluria, but unfortunately only to return to his old habit of body; for here we see the phosphates were again deposited, and continued to be so until the stone nearly reached three inches in diameter.

The occurrence of such a state of things at the present day, when the enlightened teachings of physiological chemistry have placed in our hands the means of arresting both the phosphatic and oxalic diathesis, would be a discredit to Medicine. Fortunately, however, the schoolmaster is abroad in ours as well as in other professions; and we are not likely to meet with such a case.

In the above example the oxalic was temporarily engrafted, as it were, upon the phosphatic habit of body; but this state of matters may be exactly reversed, and a person with an oxalic acid constitution may for a time suffer from the phosphatic diathesis. The truth of this statement is well illustrated by a calculus in the museum of the Royal College of Surgeons (H. b. i., described at p. 131, and figured in plate ix., of the Catalogue). The centre of the stone is composed of the oxalate of lime; surrounding this is a broad layer of white phosphates; and outside of that again is a well-marked covering of dark oxalates; thereby showing that the phosphatic only temporarily replaced the original oxalic constitution of body.

Now, just as oxaluria may be engrafted upon a phosphatic

constitution, and *vice versa*, so may oxaluria be engrafted upon the uric acid habit of body, and *vice versa*.

In 1861 a fish-dealer, aged 24, labouring under stone, applied to me for relief, and on examining the sediment in his urine with the microscope, I found that it consisted of mucus, and bladder epithelium, together with crystals of oxalate of lime, and uric acid. The coexistence of these crystals did not in the least surprise me, for I had already profited by the clinical teachings of this other calculus (No. 3662, Liston's collection, Univ. Coll. Museum), in which it is clearly shown that oxaluria may for a time be engrafted, as just said, upon the uric acid diathesis. The central part of this stone is pure, or nearly pure, uric acid ; then comes a broad layer of oxalate of lime, and lastly another layer of uric acid ; thus showing, as in the other case previously alluded to, that when the patient recovered from his attack of oxaluria it was not to get well, but only again to fall back to his old habit of body—in this instance most probably a gouty one.

As formerly said, crystals of uric acid and of oxalate of lime occasionally coexist in the same urine. The cause of this some might consider accounted for by Dr. Owen Rees's theory, which is, that not only does the uric acid diathesis frequently give rise to a deposition of oxalate of lime, but that all the oxalate of lime met with in human urine is the direct product of the decomposition of the uric acid during its sojourn in the urinary passages. He bases this assertion on the circumstance of urates being directly convertible into oxalates by boiling.

In 1859 I carefully repeated Dr. Rees's experiments on this point, but invariably with a negative result, and since then I notice that other observers, among whom may be mentioned Racle and Parkes, have equally failed in confirming the assertion. All, however, agree with that gentleman in saying that the uric acid is frequently closely associated with the oxalic

M

acid diathesis. Indeed, oxalate of lime is frequently met with as a spontaneous deposit in the urine of gouty patients, either alone or associated with uric acid, and this is not to be wondered at, seeing that many of those things which favour the appearance of oxalate of lime in the urine are precisely those which tend to create an excess of uric acid in the blood, and thereby bring on an attack of gout,—high living, excessive use of stimulants (either vinous or spirituous), sedentary habits, and such like.

The last calculus examined showed that the uric acid might both succeed and precede the oxalic acid diathesis. Here is another stone, also from Liston's collection (No. 3661 a), which shows equally clearly how the uric acid deposit may be succeeded by the deposition of oxalate of lime, and that again in its turn be followed by a deposition of phosphates. In this case, however, the deposit of phosphates may not have arisen from the patient's labouring under the phosphatic diathesis, and consequently excreting an excess of phosphates from his system, but be simply due to a local cause exciting local changes, the climax of which being the deposition of the phosphates normally existing in the urine.

These changes may be classified in three groups :—1st. The irritation caused by the presence of a stone in the bladder excites an excessive secretion of mucus, which (2nd) in its turn brings about the decomposition of the urea, the ammonia of which, by rendering the urine alkaline, (3rd) causes the precipitation of the phosphates.

This stone teaches us, therefore, another important clinical fact—namely, never to allow a calculus to remain in the bladder of a patient longer than we can help ; and if its removal is beyond our power—which is seldom the case—to prevent at least its enlargement by any subsequent depositions. The time is now arrived, in fact, when calculi ought to become exceedingly rare, for if the patient applies for advice sufficiently early the scientifically educated physician has it in his power entirely to prevent the formation of certain kinds of stone; and it is probable that the day is not far distant when a patient with a stone of any size in his bladder will be looked upon in the light of a clinical curiosity. As I said once before, the weak point in Medicine does not consist in a lack of remedies, but in a deficiency of knowledge or discrimination in the selection of them.

Even in the treatment of a case of simple oxaluria skill and judgment are required. A remedy that is suitable for one patient may be quite inappropriate for another; and even in the same patient, life being an ever fluctuating quantity, the remedy which was beneficial at one time may prove prejudicial at another. For example, I have seen cases of oxaluria which had resisted all the usual routine of mineral acid tonics, at once entirely disappear when the patient was put on the acid phosphate of soda, and *vice versa*.

In the treatment of oxaluria, as in that of every other disease, the period of life, the state of the general health, the condition of the various vital organs, the circumstances of the patient, etc., must all be taken into account if we desire to prescribe otherwise than empirically. This observation will be still better appreciated after I have called attention to the importance of the proper performance of the respiratory functions in these cases. Whatever impedes pulmonary arterialisation increases the amount of oxalic acid in the blood, and, as is known, an excess cannot long exist in the circulation without making its appearance in the urine in the form of oxalate of lime.

There can be no difficulty felt in understanding how impeded respiration gives birth to oxaluria when what was said regarding its physiology is remembered. It was then pointed out that many of the animal products, which by perfect oxidation are converted into urea and carbonic acid—among these we numbered uric acid, lactic acid, sugar, fat, creatin, leucin, tyro-sin, and guanin—by imperfect oxidation are changed into other substances, among the chief of which is oxalic acid. The lungs and kidneys are here again seen to perform a vicarious part, for when oxidation is complete the lungs exhale the carbonic acid, and when incomplete the kidneys eliminate the oxalate of lime. This, indeed, is the explanation of the so-commonly-observed fact that in the urine of patients labouring under bronchitis, phthisis, pneumonia, and emphysema, there is a temporary deposit of oxalate of lime.

Even in those diseases where the pulmonary arterialisation is only indirectly interfered with—as, for example, in heart and liver affections impeding the pulmonary circulation—oxalates likewise make their appearance in the urine.

Perhaps for an analogous reason crystals of oxalate of lime are of common occurrence in the first urine passed in cholera.

Then, again, we must never lose sight of the necessity of paying special attention to the ingesta in cases of oxaluria. Diet here has an important influence, for there can be no doubt that a large amount of oxalic acid finds its way into the system along with the food and drink. All vegetables and fruits rich in oxalic, citric, malic, and tartaric acids are to be scrupulously avoided. A little indulgence in a rhubarb tart will often cause a temporary deposition of the oxalate of lime in the urine of the most healthy, and an apple pudding greatly augment it in those predisposed to such deposits.

Champagne, Moselle, and other sparkling wines ought for a similar reason to be avoided. Even the common fermented liquors had better be left off, and if stimulants are required, the spirituous form, brandy, whisky, or gin, is to be preferred.

Carbonated alkalies are also to be refrained from, as they are transformed into oxalates in the body. The quality of the drinking water must also be inquired into, for should it by any chance contain an excess of lime, its habitual employment will completely frustrate our best efforts at treatment. If oxalic acid is to be eliminated at all from the body, it must be eliminated in a soluble form, and our remedies will utterly fail to do so if an excess of lime is at the same time present in the urine. The oxalic acid will of course combine with the lime in preference to anything else, and thereby form one of its most insoluble of salts. Lime is, in fact, the agent we employ to detect small quantities of oxalic acid in solution; so that at all hazards the lime must be got rid of from the drinking water. The simple process of boiling will separate a large quantity of it, but if that is not sufficient distilled water must be employed. This may be obtained from any neighbouring steam-engine, if nothing else is at hand; and to make perfectly sure that it is clean it may be passed through a charcoal filter before being brought to the table. As distilled water is a very unpalatable drink, a little toast may be soaked in it; or, what most patients prefer, a small quantity of brandy added.

Oxalate of lime and phosphatic calculi are the two most common forms of stone met with in districts where the water is impregnated with large quantities of lime; and it is astonishing how long it sometimes takes to arrest these deposits in persons habitually accustomed to the use of such waters. It seems as if the system in some cases required a long time to free itself of the lime with which it had become impregnated, and this is the more unfortunate as many of these are examples of renal calculi, which there is no possibility of getting at except by means of solvents.

In an early part of this lecture it was mentioned that oxaluria is most common in the prime of life; but as there are exceptions to every rule, this one is also occasionally departed from, and we meet with the oxalic acid diathesis in childhood as well as in old age. A similar remark is equally applicable to oxalate of lime calculi, which, although rarely, are still to be met with both in early and advanced life. Quite recently I saw in consultation, along with Mr. Owen, of Cleve-land-square, a gentleman who, though 70 years of age, had just passed by the urethra a small oxalate of lime calculus, which, judging from its size and appearance, must have been of very recent formation. The history in this case tended to the belief that the stone had been formed in the left kidney.

As regards the special pathology of this kind of renal calculus, one word is here necessary. For, according to Hassall, Beale, and some others, the dumb-bell-shaped crystals of oxalate of lime are more frequently the originators of mulberry renal calculi than the octahedral form. This idea, I believe, originally sprang from the fact that dumb-bell-shaped crystals are occasionally found imbedded in the fibrinous casts so commonly met with in the urine of cholera ; and as the fibrinous casts are formed in the renal tubes, the oxalate of lime crystals, it is supposed, have also formed in the same situations. Beale has even found dumb-bell crystals in the renal tubes ; but, from my own observations, I may add that not alone are dumb-bell, but also octahedral, crystals occa-sionally to be found in the tubes of the kidney ; and therefore it is highly probable that occasionally the one and occasionally the other forms the nucleus of renal mulberry gravel, and stone. It matters, however, little which it is, as the treatment of both is identical.

I cannot leave this subject without briefly calling attention to a remarkable case in which an exudation of crystals of oxalate of lime occurred on the cutaneous surface of a man labouring under retention of the urine, as it is a good illustra-tion of the vicarious action of the skin and kidneys, and forcibly points to the importance of paying attention to the one while the performance of the functions of the other are in abeyance.

The case occurred in the practice of Dr. Hingston, of Montreal. The patient, after suffering for two days from retention of urine depending upon stricture of the urethra, was attacked with phlegmasia of the left leg. The limb pitted on pressure, and patches of a bright red colour, painful to the touch, appeared about the knee. On the following day these subsided, and the patient became bathed in perspiration, with a strongly urinous odour. The face had a deep yellow cadaverous appearance, and the respiration was laboured. On the following morning—that is to say, the fourth of the reten-tion—the face and uncovered portions of the neck and hands felt to the touch as rough as sand paper. On examination the roughness was found to be due to the presence of cuboid crystals of the oxalate of lime. The patient died a day or two afterwards in a convulsion from uræmic poisoning. (a)

SULPHATES—THEIR CHEMISTRY, PHYSIOLOGY, AND PATHOLOGY.

Sulphuric acid is one of the normal constituents of human urine. It is combined with potash in the form of a soluble sulphate.

Chemistry.

Sulphuric acid needs no description. Its presence is easily detected by adding a few drops of hydrochloric acid to a drachm of urine in a test-tube, and then a drop or two of a solution of chloride of barium, when an immediate white pre-cipitate indicates the existence of sulphuric acid (sulphates). The precipitate is quite insoluble in either hydrochloric or nitric acid.

Quantitative Analysis.—Prepare a standard solution by dis-solving 30·5 grammes (473·75 grains) of dry crystalline chlo-ride of barium in 1000 cubic centimetres (32·26 oz.) of dis-

---

tilled water. One cubic cent. of this solution represents the equivalent of 0·01 gramme of sulphuric acid.

Having prepared or procured the above solution, proceed as follows :—Add to 50 c. c. of urine five drops of hydrochloric acid, and after stirring the mixture well, drop in the standard solution from the apparatus described under the head of urea until a precipitate ceases to form, or, still better, until a few drops of a solution of sulphate of magnesia, when added to a portion of the urine *plus* the standard solution, give rise to a haziness. The haziness of course indicates that all the sulphuric acid which was originally in the urine has been removed, and that there is present in the mixture an excess of the standard solution, which combines with the sulphate of magnesia. The amount in the twenty-four hours' urine is then ascertained by the ordinary method of calculation, described under the head of Urea Analysis.

Physiology.

The amount of sulphuric acid in healthy human urine depends on two things—1st, the metamorphosis of the tissues containing sulphur and sulphates ; and 2nd, on the food. The quantity daily excreted in normal urine varies from 1·123 (Becquerel and Rodier) to 2·094 grammes (Gruner, weight 120 lbs.). The minimum amount is passed during fasting; the maximum, after a full meal. Animal diet increases the amount ; vegetable food diminishes it. Exercise, both mental and physical, increases it. As only a small portion of sulphates is found in healthy blood, their use in the animal body is thought by some not to be indispensable ; thus differing very materially from the chlorides. It is generally stated, too, that there are no sulphates in the bile, gastric juice, and milk ; but this is an error, as I have detected them in all of these secretions. In the bile there is a large quantity of sulphur in the fine white crystalline product taurin. The chief part of the sulphuric acid found in the urine, however, comes from the food. Some foods are exceedingly rich in sulphur (peas, for example), and give rise to great disengagements of intestinal flatus loaded with sulphuretted hydrogen gas. Lettuce and cabbage have, in some persons, a similar effect. When alkaline sulphates are taken as medicine (sulphate of magnesia) the urine contains a large amount.

PATHOLOGY.

Bence Jones found the excretion of sulphuric acid increased in delirium tremens; and Parkes relates the case of a boy, aged 14, who, while labouring under an attack of febricula, passed the large amount of 29½ grains daily; the average amount passed by a healthy male adult on full diet being only 2 grammes, or 31 grains.

It must be admitted that there is still very little known regarding the absolute quantity of sulphuric acid eliminated in disease. All the urinary sulphates being soluble, no spontaneous precipitate of them ever takes place; and consequently but little attention is in general paid to the subject. Heller, however, says there is a great increase in the urine in inflammatory affections, even in spite of the low diet, and that, on the other hand, there is a diminution in cases of chlorosis and affections of the spinal cord, as well as in disease of the kidney. But as Lehmann has doubted the correctness of these views, the whole subject requires complete revision.

LECTURE IX.

INOSITE —INOSURIA.

INOSURIA is a term given to certain morbid conditions which
are characterised by the presence of inosite in the urine.

Chemistry.

Inosite was first discovered by Scherer in 1850 in the muscular substance of the heart, in which it exists in considerable
quantity. It is a beautiful white crystalline body, and by its
discoverer was thought to be a kind of sugar, in consequence
of its composition being represented by $C_{12}H_{12}O_{12}$, which is
precisely the same as that of sugar of milk. When analysed
without previous desiccation, it contains four equivalents of
oxygen and hydrogen more than represented above ; hence, two
equivalents of water more than cane sugar and four more than
milk sugar. Further observation has, however, shown that,
although inosite is isomeric with sugar, its reactions are too
different to allow of its being placed in the saccharine group.
For example, it neither browns with potash nor reduces the
oxide of copper, which are, as is well known, two of the most
characteristic properties of grape sugar. On the other hand, it
resembles sugar in being very soluble in water and only

sparingly soluble in alcohol; also in being by fermentation changed, among other things, into lactic acid. It is, however, incapable of alcoholic fermentation.

The presence of inosite in water is readily ascertained by adding a drop or two of nitric acid and evaporating the mixture in a capsule. After cooling, the dry residue is moistened with a drop of the chloride of lime and ammonia, and then gently warmed, when a fine rose colour indicates the presence of inosite. After a short time the colour fades away in consequence of the compound having absorbed water from the atmosphere, but on being again dried the red tint reappears.

As the presence of uric acid interferes with this test, Gallois proposes the following method for the detection of inosite in urine:—

To an ounce or two of urine a saturated solution of the neutral acetate of lead is added until a precipitate ceases to form. The mixture is then filtered, and to the clear liquid which contains the inosite basic acetate of lead solution is added, which precipitates the inosite in the form of an insoluble lead compound. After standing twenty-four hours the supernatant liquid is thrown away, and the precipitate washed with distilled water as long as any soluble matter is extracted. After this the compound is mixed with an ounce or two of distilled water and a current of sulphuretted hydrogen gas passed through it, until the lead is all separated in the form of insoluble sulphuret, which is collected on a filter, while the clear filtrate containing the inosite is evaporated to nearly dryness. When nearly dry, a drop of the nitric oxide of mercury is added, heat is again applied, and as the liquid becomes dry a fine rose colour is produced if inosite be present.

When albuminous urine is being tested, the albumen must be first separated with acetic acid and heat. In the case of diabetic urine, on the other hand, tribasic instead of the ordinary basic acetate of lead solution is employed, and the

precipitate washed so long as sugar comes away, which is easily
ascertained by the ordinary reactions.   In other respects the
method of analysis is the same.

Physiology.

As already said, Scherer found inosite in the tissue of the
heart ; but this is not the only organ of the body in which it
is met with.   In 1852, while working with Robin and Verdeil,
I discovered inosite in the voluntary muscular fibre of the ox,
and succeeded in preparing some very fine crystals of its compound with baryta from a few pounds of beef-steak.

An admirable drawing, as well as a full description of these
crystals, was subsequently given by Robin and Verdeil in their
" Chimie Anatomique." (a)   The subjoined figure (16) is a

FIG. 16.

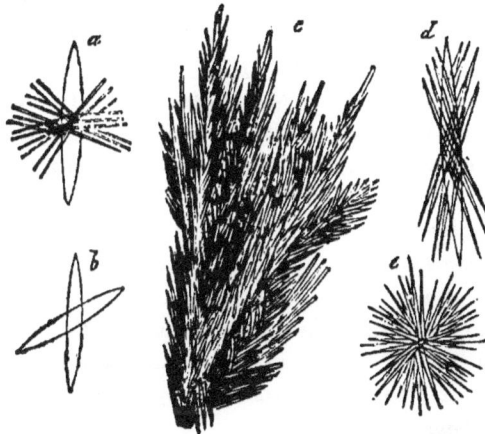

Crystals of the inosite of baryta from voluntary muscle.

copy of a few of the most characteristic forms there represented.
The crystals are colourless and transparent.   Some appear as
groups of fine needles (*d*), others as stars (*e*), and a few of the
larger bundles are fan-shaped (*c*).   Occasionally also oblong

---

(a) *Vide* Vol. II., p. 459, and Plate XLV. of the Atlas.

plates, singly or in groups (*a, b*), are found mixed up with a few free needles.

Inosite is not alone to be met with in the muscular tissues of the organism, but also in the brain (Müller), lungs (Cloetta), spleen, and pancreas. It must, therefore, be regarded as one of the normal constituents of the body. The part it plays there is, however, still unknown, although, in consequence of its crystalline nature, it is highly probable that it is—like urea and uric acid—one of the effete products of healthy action.

Pathology.

According to the recent researches of Dr. N. Gallois, (a) healthy human urine is devoid of inosite, whereas, as Cloetta first pointed out, both in the urine of albuminuria and diabetes, it exists in appreciable quantity.

Although it would thus appear that both sugar and inosite may present themselves in the urine at the same time, they are, nevertheless, more generally observed separately than combined.

The presence of inosite in the urine *per se*—that is, without being associated with any other urinary affection—such, for example, as albuminuria or diabetes, can scarcely be said to constitute a known disease ; for inosuria, as it has been called, has not as yet been found to possess any characteristic symptoms, all that is known on the subject being that perfectly normal urine contains none.

As it is not at all unlikely that at no very distant date this substance may assume a clinical importance, it may be useful briefly to mention a few of the pathological conditions in which it has already been met with. The following are extracted from J. Neukomm's Thesis (Zurich, 1859). The researches were made in Staedeler's laboratory, and consequently under his supervision.

(a) Monograph "De L'Inosurie." Paris. 1864.

Bright's Disease.—Male, aged 34 ; six months ill ; albuminuria, œdema, convulsions, and coma. The urine during
life contained a small quantity of inosite, and after death the
brain was found to contain abundance of that substance.

Diabetes.—Man, aged 29 ; the progress of the disease during
the latter period was very rapid. After death the kidneys
were found to contain only a small quantity of inosite, while
the brain contained it in abundance.

Syphilitic cachexia.—In which there was ascites and œdema
of the legs with a rapid diminution of strength, the patient
dying on the sixth day. The kidneys contained a large quantity of inosite, whereas the brain gave evidence of none, exactly
the reverse of what was found in the preceding two cases.

Typhus.—In a young man, who died in the third week ; no
inosite could be found either in the heart or kidneys.

Lastly, in a case of Phthisis in a girl, aged 17, a remarkable
quantity of inosite was found in the brain, and scarcely any
in the heart.

These cases are far too few to admit of our drawing any
conclusions from them, except it may, perhaps, be that the
proportion of inosite present in the organs during health and
disease is occasionally reversed. Gallois, who has paid considerable attention to the subject, thinks it probable that an
excess of inosite in the urine arises from the imperfect performance of the glucogenic function of the liver, for even in
cases of artificial diabetes in animals (produced by puncturing
the fourth ventricle of the brain) he occasionally found inosuria
accidentally produced.

CREATIN AND CREATININ IN HEALTH AND DISEASE.

In the animal body we meet with two substances, which
are so closely allied to each other that they have received the
respective names of creatin and creatinin ; and although it is

only the latter that is excreted from the body along with the urine, the former, in consequence of the ready convertibility of one substance into the other, is with almost equal frequency encountered in the renal secretion. In order, therefore, clearly to explain this subject it will be necessary for me briefly to mention the chemical characters and physiological history of each.

Chemistry.

Creatin is a white crystalline organic substance, with a neutral reaction. It is soluble in water, slightly soluble in alcohol, and insoluble in ether. When boiled with strong alkalies it is decomposed, and converted into urea, sarcosine, and, if the boiling be continued sufficiently long, into carbonate of ammonia.

When heated with strong acids it is transformed into *creatinin*, a substance closely allied to it, the only difference in their composition being that it contains two equivalents less of water than creatin. In properties, on the other hand, the substances are very dissimilar, creatinin being alkaline, and not only soluble in water, but in alcohol also, and even slightly soluble in ether.

Both creatin and creatinin may be separated from the urine by adding to it milk of lime and chloride of calcium, filtering the liquid from the precipitate, then concentrating it to a syrup and adding a little chloride of zinc, which combines with the creatin and creatinin to form a crystalline compound easily separable by further filtration. This crystalline compound has next to be decomposed by dissolving it in warm water, and adding a small quantity of hydrated oxide of lead. The liquid has again to be filtered from the precipitate of oxide of zinc and lead, and then boiled with charcoal to remove the colouring matter, again filtered, and again concentrated and allowed to stand until the creatin and creatinin crystallize. The aqueous solution of creatinin gives a flocculent white precipitate with the bichloride of mercury, and reddish crystals with the bichloride of platinum.

Physiology.

Creatin was first discovered in the juice of flesh by Liebig, in the blood by Verdeil and Marcet, and in the urine by Heintz ; and although creatinin has also been detected in all of these situations (Liebig), there is this remarkable difference between the two substances, that while creatin is most abundant in the tissues, creatinin occurs in the largest quantity in the urine. This fact has formed the foundation of the theory that creatin is the direct product of muscular action, and creatinin the result of the still further retrograde metamorphosis of the creatin. This view is supposed to gain additional support from the circumstance that by the simple act of decomposition all the creatin present in urine is speedily converted into creatinin. Some say that creatin is normally excreted by the kidneys as well as creatinin, but this is a point upon which further evidence is required. Meanwhile, it may be mentioned that the respective amounts of these substances said to be excreted during the twenty-four hours by healthy adults is 0·305 gramme (4·7 grains) of creatin (Thudichum), and 1 gramme (15½ grains) of creatinin (Neubauer). One curious fact connected with these substances is that, under certain conditions, the creatinin is reconvertible into creatin. Thus, on one occasion Liebig found in the urine of a dog, which had been fed solely on flesh, a great quantity of creatinin, and no creatin whatever ; whereas after this urine had been treated with milk of lime, and allowed to stand for six weeks, abundance of creatin was detectable. It is presumed, therefore, that the lime had in this case reconverted the creatinin into creatin.

Pathology.

These substances so very rarely occur as spontaneous deposits that little is at present known regarding their pathological importance. In March, 1861, I injected into the portal vein of a small dog three drops of alcohol in two drachms of water. Two days afterwards the urine contained a small quantity of sugar, and in two days more the dog was sacrificed, and his urine removed from the bladder. It was then clear and tolerably high coloured, but was no longer saccharine. It was put aside, and in ten days the surface of the liquid, which was still very acid, was covered with large crystals, which were at first thought to be phosphates.

FIG. 17.

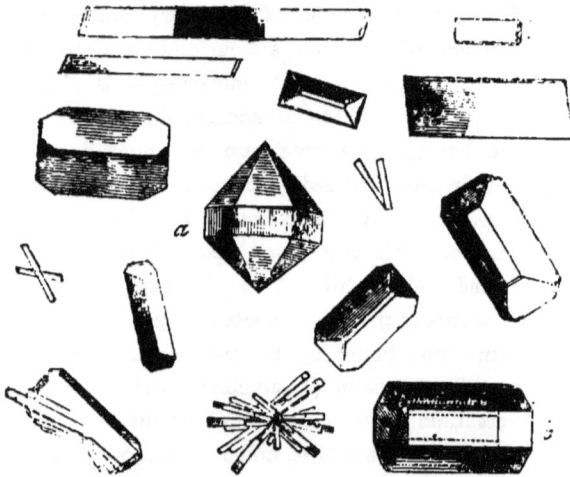

FIG. 17.—Crystals of creatin, after Robin and Verdeil.

Two circumstances, however, raised a doubt in my mind as to their nature—first, the fact of the urine having an acid reaction; and second, the enormous size of the crystals. The bottom of the vessel also contained a large quantity, so there was no difficulty found in further investigating them. They were ascertained to be beautiful polariscopic objects of an organic nature, and of the forms represented in the above figure (17).

N

On adding nitric acid to some of the urine just as it stood, it was so loaded with urea that the whole became one mass of crystals. Since making the above observation, I have frequently set aside urines with the view of again obtaining a similar spontaneous deposit, and I have seen it once in a case of diabetes in a young clergyman, twice again in dogs fed on flesh diet, and rendered artificially diabetic, and once in the urine of a stall-fed London cow. The cow's urine is normally alkaline; but in this case it was acid even after it had stood four days (November 4, 1862). When passed it was very watery; and all that was done, was merely to evaporate it to one-half over the water bath. The crystals were very large, being not only visible to the naked eye, but some of them quite big enough to be picked up with the fingers.

CHOLESTERIN AS A URINARY DEPOSIT.

Chemistry.

Cholesterin is a beautiful white crystalline substance, in many respects resembling a fat—indeed, for several years it went under the name of "biliary fat." It is quite insoluble in water, alkalies, and acids; very soluble in alcohol and ether; highly inflammable, and destitute of nitrogen; its composition in 100 parts being—

Carbon	84·00
Hydrogen	12·00
Oxygen	4·00
	100·00

It is readily distinguished from all ordinary fatty matters in consequence of being non-saponifiable with alkalies. When pure, cholesterin is found in thin, white, transparent, rhom-

boidal plates, with parallel sides and sharp angles. As represented in fig. 18, the crystals are occasionally large and
superimposed on one another, but in consequence of their
transparency the outlines of the under strata are as well seen
as those of the upper. Hence they are fine microscopic
objects, visible with a low power.

The easiest test for cholesterin is its microscopic characters,
in combination with its extreme solubility in hot alcohol, and
re-deposition on cooling in a crystalline form. If the cholesterin
be in a dry state it gives along with sulphuric acid on a white
porcelain vessel a beautiful play of colours, passing through
the different shades of orange, red, purple, and green.

FIG. 18.

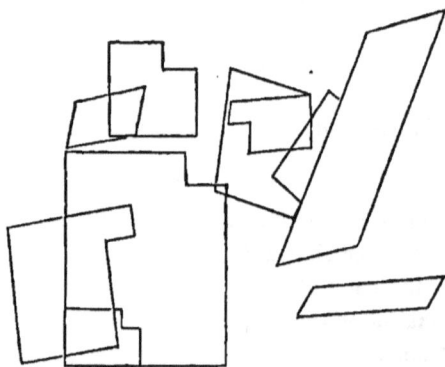

FIG. 18.—Crystals of Cholesterin spontaneously formed.

By prolonged boiling with nitric acid cholesterin is converted into (among other things) oxalic, acetic, and butyric
acids.

Physiology.

Cholesterin is one of the normal ingredients of the human
body. It is met with in large quantity in the bile, in which
it is retained in a soluble state by glycocholic and taurocholic

acids, in the aqueous solutions of which it is exceedingly soluble. Cholesterin is also a normal constituent of the nervous system, being very abundant in the brain and spinal cord. Moreover, it is to be met with in the tissue of the liver and spleen, and is also regarded in the light of a constant constituent of the blood.

Cholesterin seems to undergo decomposition in the body, for as far as I am aware it has never been found as a product of any of the normal excretions, although it is abundantly to be met with in a great variety of situations during disease.

Pathology.

Cholesterin was originally discovered in biliary calculi; many—indeed, I might say most, of which are almost entirely composed of this substance. On a few rare occasions the calculus may be said to consist of pure cholesterin. Here is one, from my own collection, which is, as you perceive, of almost transparent purity, notwithstanding that it is as large as a pigeon's egg.

Cholesterin is a common constituent of liquid exudations, being found in large quantity in the fluid of ovarian cysts, hydroceles, and hydatid cysts.

It is also to be constantly met with in the pasty-like matter of sebaceous cysts ; in fact, there are few cystic tumours, be their contents solid or liquid, to be found without it.

Lehmann has also met with crystals of cholesterin both in the choroid plexus of the brain and in the pulmonary expectoration of advanced phthisis.

In calcified tubercle and echinococci it is frequently to be met with.

Notwithstanding it being such a very common pathological product, cholesterin was not until lately believed to be excreted with the urine. Müller was the first to call attention to its presence in the renal excretion, having on two occasions met with crystals of it in the kiestein of pregnancy. Gmelin also found cholesterin in the urine in a case of jaundice from obstructed bile-duct; and since then various authors have called attention to its occasional presence in the renal secretion. To Dr. Beale we are chiefly indebted, however, for giving prominence to the subject. He detected it in the urine in several cases of fatty degeneration of the kidneys in considerable quantity, and also called attention to the important fact that the oil globules found at the bottom of the vessel in cases of kidney disease consist, not of the ordinary fats, as hitherto supposed, but of cholesterin, as can be proved, he says, by dissolving them in alcohol and recrystallising. Cholesterin thus differs from other fatty substances in being heavier than water.

It may, perhaps, not be out of place to mention the fact that cholesterin in a crystalline form is not unfrequently met with in pus, more especially after it has stood a few days.

The presence of cholesterin in one and all of the before-mentioned abnormal situations, is not, however, of any great clinical import except when it appears in the sputa or in the urine; for then, and then alone, can it be regarded as a sign of severely-disordered tissue metamorphosis.

Cystin as a Urinary Deposit, and as Calculi—Its Nature and Treatment.

Cystin, which was first discovered in 1805 by Dr. Wollaston in a urinary calculus, enjoys the remarkable peculiarity of consisting of no less than a quarter of its weight of pure sulphur.

Chemistry.

Before exposure to light and air, Cystin is a pale, yellow,

opaque, amber-coloured crystalline organic substance, devoid
of either smell, taste, or reaction. In 100 parts it is composed of

Carbon	30
SULPHUR.	27
Oxygen	26
Nitrogen	12
Hydrogen	5
	100

After prolonged exposure to light and air it changes its
colour, and becomes of a sea-green hue. It has several peculiar
chemical characters, being soluble in ammonia, but insoluble
in carbonate of ammonia; soluble in oxalic, but insoluble in
acetic and tartaric acids; soluble both in the mineral acids
and alkalies, yet insoluble in water and alcohol. Moreover,
cystin combines with the mineral acids to form salts, whereas
the alkalies merely dissolve it.

From its acid solutions it is precipitated by the carbonate of
ammonia, and from its alkaline solutions by acetic acid.
When boiled with caustic potash, it yields ammonia and an
inflammable gas.

When strongly heated in the open air it does not fuse,
but burns with a bluish-green flame, and gives off an
acid vapour with a disagreeable odour. Whereas, when
heated in a close vessel (dry distillation), it chars, yields a
fetid oil, and gives off the alkaline vapours of ammonia.

Cystin is readily detected in the urine in consequence of its
possessing well-defined crystalline characters. Under the
microscope the crystals are seen either to be perfect hexa-
gons or some modification of the hexagon (Fig. 18, a).
They are thin plates, and often lie one over the other (c).
So characteristic is their appearance that it is almost impossible to confound them with any other urinary deposit, unless

it might perhaps be with uric acid, which, on some rare occasions, crystallizes in hexagonal-looking plates. The cystin crystals are, however, much smaller than those of uric acid, and even when there is a doubt about the nature of the deposit the addition of a few drops of a mineral acid soon settles the question, as the crystals of uric acid remain unaffected, while those of cystin are speedily dissolved, again to reappear on the neutralisation of the acid by the carbonate of ammonia.

FIG. 18.

FIG. 18.—Cystin. *a* represents crystals spontaneously formed in human urine; *b*, deposited after the addition of acetic acid; *c*, crystals obtained by dissolving a portion of a calculus in ammonia and evaporating to dryness.

It must be borne in mind that crystals are not always to be detected in urine containing cystin. For there are many cases in which the urine is alkaline, or, at least, neutral, where no deposit is to be detected until after the addition of acetic acid. The cystin is then thrown down either as an amorphous precipitate or in the shape of imperfectly formed crystals (Fig. 18, *b*). The latter are occasionally not unlike balls of tyrosin, appearing in some cases to consist of badly formed rosettes of small crystals.

Should a case arise where the deposit, caused by acetic acid, is likely to be confounded with urates, strong nitric acid ought to be added, which will at once dissolve the cystin, and leave the other intact. If still further proof of the true nature of the deposit be required, collect, dry, and redissolve it in ammonia, and on the slow evaporation of the ammonia the cystin will assume the hexagonal form of crystallisation.

Urine containing a larger quantity of cystin has in general a pale greenish yellow hue, and a sweet briar-like odour. The cause of the odour I am at some loss to comprehend, seeing that pure cystin is perfectly odourless. When stale, cystin urine smells like rotten eggs, in consequence of its sulphur being transformed into sulphuretted hydrogen.

In testing calculi for cystin, place a portion of the stone on platinum foil and burn it. The following characters will denote the presence of cystin :—1st, the sickening acid smell ; 2nd, the bluish green flame ; and, 3rd (what I consider a very characteristic test), the polished platinum soon becomes stained of a deep greenish blue colour, which colour entirely disappears on further heating.

If the stone be impure the quantity of cystin it contains can be readily ascertained by dissolving a given amount in ammonia, filtering and weighing the cystin left after the evaporation of the liquid.

Physiology.

The origin of cystin in the animal body is as yet entirely unknown, but there are grounds for supposing that it is in some way or other connected with the function of the liver. The first of these is that the only substance of a crystalline nature met with in the healthy body containing sulphur is taurin, one of the normal products of the bile. The second, that in certain diseased states of the system crystals of cystin are met with in the liver (Scherer, Cloetta, Virchow, and myself).

At one time cystin was, like many of the other urinary ingredients, supposed to be formed in the kidneys, but the fact of its being occasionally met with in the liver completely negatives that view. Indeed, all recent research tends to prove that the kidneys are in no case formative organs, and that their function is purely excretive.

It is worthy of observation that taurin, the substance supposed to yield cystin, contains an almost equal per centage of sulphur, namely, 25 per cent. As the physiology of cystin deposits is still shrouded in obscurity, and the very existence of it in the healthy animal body is denied, we may at once pass on to its

Pathology.

Notwithstanding that cystin is never to be met with in the healthy animal organism, its presence in disease is by no means so rare as one might at first sight imagine. Indeed, it is highly probable that it is often present in the urine without being detected, even in cases where the secretion has been microscopically examined. I have two reasons for saying so ; first, cystin is often retained in solution until it is artificially precipitated by means of acetic acid ; and secondly, the cystin diathesis has in most cases remained undiscovered not only until a stone has formed, but even until the nature of the calculus has been revealed by chemical analysis.

Although the presence of cystin in the urine must be regarded as abnormal, yet it cannot be looked upon as an indication of the existence of any grave disease, for until it accumulates in the kidney or bladder and gives rise to the discomforts of a stone, the persons labouring under it do not appear to be in worse health than the majority of their neighbours. An old fellow student of my own has to my personal knowledge had cystin deposit in his urine for at least fifteen or sixteen years, and not until very recently did it cause him any discomfort (see the postscript).

The cystin diathesis has one striking peculiarity, namely, that it is seldom preceded, associated with, or followed by any other. Indeed, persons labouring under it seem to enjoy a remarkable immunity from other urinary deposits. Look, for example, at these three fine specimens of cystin calculi; how different they are from many of the stones we have already examined. They have neither nuclei of uric acid, nor coatings of phosphates. From beginning to end they are uniform in structure, and these are no exceptions to the general rule, for although I have examined twelve of this species of calculus, not one of them had either a nucleus or a covering of any other kind of urinary sediment, and only in very few of them was the cystin, as in one of the specimens before us, mixed with an impurity. Even in glancing over the literature of the subject I have succeeded in finding only two cases in which the calculus was of a complex character. The first is one in which a cystin stone had a uric acid covering (recorded by Henry), and the second is one in which a cystin stone had a uric acid nucleus (mentioned by Yelloly).

The urinary deposit which appears to have the greatest tendency to mix itself up with cystin is that of phosphates, and it seems to me highly probable that this is not in any way due to the patient having a phosphatic diathesis, but merely to the accidental circumstance of cystin urine frequently having an alkaline reaction. Phosphates being, as is well known, insoluble in alkaline urine, we have at once a good chemical reason for their occasionally becoming mixed up with cystin deposits. A precisely similar remark is equally applicable to the sparingly soluble carbonates and oxalates of the alkaline earths.

The cystin is easily separable from the phosphatic deposit by treating the mixture with strong acetic acid, which, while it dissolves the phosphates, only renders the cystin more insoluble.

If the word diathesis is applicable to anything at all, it seems specially appropriate to those cases in which a cystin deposit is habitually met with in the human urine; for if we examine into the history of the recorded cases, we find it impossible to arrive at any other conclusion than that the cystin deposit depends upon a peculiar congenital or acquired state of body, quite independent of any special disease or known constitutional peculiarity.

The majority of writers, I am aware, speak of it as being hereditary, in consequence of it having several times been found to occur in more than one member of the same family. Civiale, for example, met with it in two brothers; Toel in two sisters; and Golding Bird in no less than three succeeding generations. Notwithstanding this, however, I am disinclined to adopt the hereditary theory; for on looking through the numerous recorded cases of cystin calculi, it will be found that in very many instances special mention is made that the patient was the only member of his family known to be so afflicted.

On the other hand again, cystin deposits cannot be said to depend on national peculiarities, for England, Scotland, France, Ireland, and Germany have all furnished their quota. Neither can we localise them in any particular part of the country; nor can we attribute them to any special kind of drink; for cystin calculi are alike to be met with in cider-drinking Herefordshire, beer-consuming Yorkshire, and whisky-imbibing Mid Lothian. Some think that fish and vegetable diet have a great effect in producing them; but they have utterly failed to prove that either the pisciverous Shetlander, or the modern vegetarian are more liable to cystin calculi than persons who indulge in a less restricted form of diet.

Golding Bird hazarded the idea that scrofulous persons were more liable to the cystin diathesis than others—a doctrine which appears to me as untenable as that which attributes it to the chlorotic habit of body. A glance at the thousands of persons surrounding us who suffer from the one or other of these conditions, and in whose urine no cystin deposit has ever been detected, is sufficient of itself to negative either of these views.

FIG. 19.

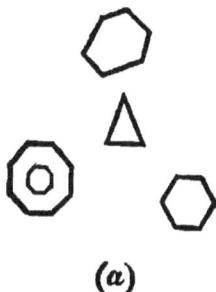

(a)

FIG. 19.—Crystals of cystin from a diseased human liver. ·

The cystin diathesis, be it what it may, is certainly something apart from any other condition of body with which we are familiar, and, as said before, the most tenable doctrine appears to be that it is in some way or other connected with the interrupted metamorphosis of some of the hepatic products, as they alone possess the required amount of sulphur, and in the urine only do crystals of cystin appear as a common deposit. The accompanying figure (19) represents some crystals of cystin which I detected in the liver in a fatal case of jaundice. (b)

It would seem, too, that the habit of body in which it originates is more likely to be congenital than acquired, for among the cases recorded the majority bear evidence that the cystin deposit began at a very early age, and not one is to be found in which it was met with after the prime of life.

It may be well here to remark that in some cases the deposition of cystin in the urine is irregularly intermittent. In the case of a patient recovering from a severe attack of pneumonia, who was under my care in 1850, while Resident Physician in the Royal Infirmary, Edinburgh, the urine contained well-formed crystals of cystin every second or third day. A connexion between the diet and the deposit in this case was sought to be discovered, but with an entirely negative result, except on one point, which appeared to be pretty satisfactorily established—namely, that as the disease of the

---

(b) A complete history of this case will be found in the author's work on "Jaundice and Diseases of the Liver and Pancreas," p. 86.

lungs disappeared, and the appetite of the patient improved, the amount of cystin in the urine gradually increased.

As regards treatment, it is much to be regretted that no rules can as yet be given which have satisfactorily stood the test of experience ; but grounding our views on a knowledge of the chemical composition and reactions of cystin, I would venture to advise the adoption of the following line of treatment :—

1st. As to diet, all foods rich in sulphur ought to be avoided. Eggs, peas, beans, and cabbage would of course come under this head. Protein substances—albumen and fibrin, for example—which also contain sulphur, should be taken in moderation ; whereas, the numerous articles of diet belonging to the starchy, saccharine, and oleaginous groups may be indulged in with impunity. For similar reasons all drinks containing sulphates ought to be shunned. This rule will, therefore, necessitate the discontinuance of the use of the majority of mineral waters, especially the purgative ones, which are in general rich in sulphates. Pickles ought also to be refrained from in consequence of the acetic acid they contain ; cystin being insoluble in that acid.

2nd. With respect to remedies, for the reason just given, sulphate of magnesia is, of course, not to be employed as a purgative, and if a mineral acid is to be administered, one would naturally refrain from selecting sulphuric for that purpose ; as both of these remedies would only add fuel to the fire. Then, again, as cystin is insoluble in acetic and tartaric acids, if our patient is threatened with a stone we would prevent him indulging in substances likely to contain either the one or other of these acids, which even although converted in the system into carbonic acid would still be objectionable ; while we would at the same time avoid prescribing for him either citrates, acetates, or tartrates. For similar reasons an embargo should be put upon the carbonate of ammonia, and even ammonia in any form, as it is speedily transformed into the carbonate in the circulation.

The other alkalies had also better be avoided, not because cystin is insoluble in them, but because the urine in persons affected with the cystin diathesis has a tendency to alkalinity ; and in alkaline urine there is always a danger of a deposition of phosphates, oxalates, or carbonates taking place.

Ferruginous tonics have been greatly recommended in the cases we are now considering ; and no doubt whatever tends to restore the healthy tone to the system is likely to do good. If, however, there be one form of remedy more likely to benefit than another, it appears to me that it is that of the mineral acids, nitric and hydrochloric (not sulphuric), as they will not only help to retain the cystin in solution, but directly aid in correcting the tendency to alkalinity in the blood and urine. These mineral acids can, of course, easily be combined with vegetable tonics, and, if need be, with several ferruginous ones.

Some, not yet emancipated from the fetters of the so-called "practical" delusion, may be tempted to exclaim against this as being chemical therapeutics ; but to such I reply, that I am proud to say that it is chemical therapeutics ; and that if chemical therapeutics are worth anything, they are most assuredly so in the treatment of diseases of the urinary organs. All the blundering empiricism of past times might have in great part been avoided, had our predecessors only known a little more of the chemistry of the body and its secretions.

A few years ago scientific farming was looked upon as a wild illusion, and the so-called "practical" wiseacres shook their wise heads when Liebig, in propounding to them the self-evident doctrine that out of nothing, nothing can be made, pointed out that, in order to procure crops, the inorganic and organic substances comprising them must be furnished by the

soil, supplied in the water, or yielded from the atmosphere.
Until then the seed, in the eyes of the farmer, was beyond the
pale of any natural law, and growth was by him regarded as
a mysterious process by which the plant, under the influence
of heat, light, and moisture, manufactured all that it
required out of an unknown something, about the nature
of which few apparently sought to inquire. Now, however, the light of knowledge is gradually dawning upon
the agriculturist, and he begins to discover that plants,
although able to build up, transform, arrange, and modify,
can neither create nor destroy a single atom of the elements
of which they are made ; and although he still pretends
to scorn Liebig and his doctrines, he cannot help perceiv-
ing that the fertility of his fields increases and decreases, other
things being equal, in a direct ratio to the amount of food they
contain suitable to each particular crop. Nay, more, he even
begins to try and restore to the exhausted soil, by means of
artificial manures, those substances of which he supposes it
has become deficient. His ignorance is still, however, such
that, as soon as he fails in his endeavour to restore fertility
by artificial products, he lays the blame at the door of
chemistry, and entirely overlooks the fact that it arose from
his own error in not adding the proper ingredient in the proper
place ; for should his fields have become exhausted of phosphates, and he manures them with ammonia, which they
already possess in superabundance, is it surprising that he
obtains a stunted crop of a vegetable whose tissues require
more phosphatic than ammoniacal salts ? Every one, of
course, exclaims no ; but why, then, I ask, do we not apply
the same mode of reasoning and the same remark to our own
failures in medicine ? In what respect do we differ from the
agriculturist ? His object is to obtain good crops, and so is
ours ; the only difference being that while he looks for corn
and wheat, we seek to obtain muscular strength and mental

vigour, and the animal body can no more generate these out of nothing, than the vegetable can fruit and flowers.

Just as there is a fundamental law in physics, that no matter can exist without force, and no force can be generated without matter ; so, too, in physiology, there is a fundamental law, that no change can take place in the material of which the animal body is composed without giving rise to vital phenomena, and no vital phenomena can take place without a corresponding change in the material. And as the animal body, like the vegetable, can neither create nor destroy a single element—in order that it may produce muscular strength and mental power, a constant supply of the proper material must be furnished to it; and, in cases where the food is deficient in the required amount of material, we must add it in the shape of medicine.

In giving medicine, we but supply directly or indirectly to the different tissues of the frame the materials of which they have become exhausted, or hasten the removal of those which are in excess or have become effete ;—and the sooner practical men become alive to the truth of this all-important doctrine, the better it will be both for themselves and their patients. I am perfectly aware that this "material" view of therapeutics, like Liebig's views of agriculture, will be scouted at for a season ; but, being thoroughly convinced of the stability of the foundation upon which it is based, I have no fear of the ultimate result, and am perfectly content to bide my time.

Let us look for a moment at the question of cystin. No one can deny that it contains a large quantity of sulphur, and I think all will admit that as sulphur is an " element," it cannot be manufactured in the body ; for although the animal organism can form innumerable compounds out of the fourteen elements of which it is composed, it cannot create the smallest particle of one of the fourteen elements themselves. The sulphur contained in the eliminated cystin therefore, being obtained by the body from without, it is clear that by stopping the supply of this element we could at once arrest the formation of cystin. As sulphur is one of those substances, however, which performs a multitude of important offices in the animal body, it would never do entirely to stop its supply. The next best thing, therefore, is to diminish it to the minimum which the wants of the system demand, and to try and remove by other means the abnormal tendency the body has of combining it with O. H. and N. in the form of an abnormal compound.

The line of treatment which has been above suggested is that which at present appears to be the best, but I do not for a moment pretend to say that it is absolutely so, or that it will not require considerable modification according to the circumstances and constitutional peculiarity of the patient. The difficulties in the application of chemical therapeutics are neither less nor more than those which stand in the way of empirical therapeutics; for in either case, if we even take for granted that our diagnosis is perfectly correct, and that the remedy selected is the most suitable contained in the Pharmacopœia, we can never foretell what extraneous influences may be brought to bear upon the body and modify or even counteract its effects. This much, however, may be said, that as no two cases are ever precisely alike, and no two sets of conditions invariably the same, there is far less likelihood of error in rational than in empirical medicine; for while the former combines the teachings of science with those of empiricism, the latter contents itself with the information obtained from empiricism alone.

The chief obstacle to the spread of rational, or as they are often called, chemical therapeutics, lies in the imperfect knowledge of physiology and chemistry possessed by practical men. For the limited acquaintance they have with these studies

o

prevents their appreciating the true value of scientific information sufficiently to deter them from attempting to apply it
in practice ere they have thoroughly mastered the subject.
The natural result consequently is that they almost invariably fail in their endeavours to combine science with empiricism, but instead of attributing their failure to a deficiency
of knowledge, put the entire blame upon what they please
to term the false teachings of science. It is to be hoped,
however, that through the agency of physiological laboratories, such as have been already established in France and
in Germany, at this (University) College and at Netley,
the succeeding generations of Medical men will be enabled
more successfully to apply these same teachings of science
to the alleviation of suffering, and thereby raise our Profession
to the high position it is evidently destined at no very distant
period to occupy.

As regards cystin calculi, some special remarks are necessary. They are by no means always vesical; for Marcet,
senior, mentions no less than three cases in which they were
met with in the kidneys ; and Berzelius, Brande, and Bence
Jones have each added one to this number. The occurrence
of renal cystin calculi of course proves that, under certain
conditions of the system, the state of the urine is such, even
at the moment of its secretion, as to prevent it retaining the
cystin in solution,—a fact of considerable clinical importance ;
and it would be exceedingly useful to ascertain if, in cases of
renal cystin calculi, the urine is alkaline ; for if so, the necessity of the acid line of treatment would be made more
apparent.

Cystin calculi may form at any period of life, from birth
onwards to middle age. Wollaston's celebrated calculus was
removed from the urethra of an infant, and Müller has
described one taken from the bladder of a boy six years of age.
The majority appear to have been met with between the

twentieth and thirtieth year, although their formation seems
to have begun much earlier.

It has already been mentioned that cystin calculi are seldom
of a complex character, and it may now be added that they
are even very rarely of a mixed nature. Here is one (No. 2457,
Univ. Coll. Museum) which well illustrates how beautifully pure they may sometimes be. Scrapings from this stone
almost entirely dissolve in ammonia, and on the evaporation of
the liquid fine hexagonal plates of cystin cover the bottom
and sides of the vessel. When analysing this specimen a few
years ago, I had occasion to incinerate a portion, and it was so
pure as to be entirely dissipated by heat. No inorganic ash
remained behind. This half of the stone in its present state
weighs 265 grains, so that the entire specimen, before it was
cut and polished, must have weighed considerably more than
530 grains.

The effect of light on cystin is well shown in this specimen. All the exposed parts are of a fine sea-green colour,
while that in contact with the cotton wool is of a brownish-
yellow amber hue. Moreover, the polished section further
shows the confusedly crystalline-looking structure presented
by pure cystin, which is of itself a sign of its purity. The
outside of the stone is rough, tuberculated, and glistening
from the peculiar arrangement of the crystals. In shape the
stone is a flattened oval, the surface of the section being $2\frac{1}{3}$ by
$1\frac{1}{4}$ inches in diameter. The only history connected with the
specimen is that it was successfully removed by Mr. Liston
from the bladder of a Scotchman, aged 37, none of whose
relations were known to have suffered from stone.

This other specimen (No. 3620) is one of equal interest, but
in another way. It consists, as is seen, of a number of fragments, amounting in all to above half-an-ounce in weight,
which were passed by the urethra after the calculus was
crushed by Mr. Liston. The patient was a young man, 19

years of age, a native of Herefordshire, and the symptoms of
stone had existed for many years before he came to University
College Hospital. The operation of crushing was repeated
four times before all the fragments came away. Amongst
them is a small oval calculus of the same material, with a fine
crystalline exterior. It was passed along with the fragments.
It has more of an amber yellow than of a green colour, which
Dr. Prout, who examined the specimens, considered might
be due to it containing some more of the water of crystallisation than the larger stone ; but it appears to me to be chiefly
due to its being kept covered up in paper, while the others
are exposed to the light. No hereditary taint could be made
out in this case, and no other member of the family was
known to have suffered from stone of any kind.

This third specimen of cystin calculi (No. 2621, Univ. Col.
Museum) is also exceedingly interesting, notwithstanding that
nothing is known of the history of the patient from whom it
was removed. The stone, however, tells part of its own
clinical history, and a most instructive part, as we shall
immediately see.

In the first place, the section of this stone is not of the same
colour as that of the other specimens : it is of a much paler
green, looking more like tailors' chalk than anything else.
Besides this, it has not the confusedly crystalline appearance, nor the transparency so characteristic of pure cystin.
Whereas, it possesses, on the other hand, a distinctly, though
slightly, striated arrangement, which I consider of itself an
indication of impurity. In this case the impurity consists of
an admixture of phosphates, and these are so uniformly blended
with the cystin that I am led to the belief that the presence
of phosphate of lime in this instance is not due to the patient
having had a phosphatic as well as a cystin diathesis, but solely
to the accidental circumstance of his urine having been alkaline, which would, of course, of itself fully account for the

phosphatic admixture. The stone in its present dilapidated condition weighs nearly 850 grains, and is two and a quarter inches in its greatest diameter(a).

It was previously said that many patients labouring under the cystin diathesis do not appear to enjoy less health than their neighbours until they begin to suffer from symptoms of stone. It may now be added, that cystin calculi do not appear to give rise to so much discomfort as many other kinds of stone. But whether or not this arises from the general health of the patients being better I am not in a position to say, the statement being merely based on the two following facts :—

1stly. Most cystin calculi are of considerable size before the patient applies for relief, and 2ndly, there is an almost invariable absence of any phosphatic covering on cystin calculi, which proves that they do not give rise to the same amount of vesical irritation as other forms of stone, the majority of which, when of any size, have a coating of phosphates.

This leads me to mention that I have met with two cases of cerebral disturbance, in which cystin appeared in the urine, associated with uric acid. The first was that of a gentleman, aged 43, whom I several times saw, in consultation with Mr. William Coulson, during the latter part of 1869, and beginning of 1870. The second was that of a lady aged 53, sent to me in May, 1870, by Professor Attfield, who discovered cystin along with uric acid in her urine. Both patients were subject to gravel, both of a highly nervous temperament, the former with symptoms of derangement of the spine, the latter of the brain. In Mr. William Coulson's case the uric acid was in very large, the cystin only in small quantity. The uric acid resisted, the cystin yielded to the treatment. In the case of the lady, the uric acid and cystin were nearly in equal proportions, and equally amenable to treatment.

P.S —As it is not often that one hears a medical man's views

(a) Dr. B. Jones published an account of the analysis of this calculus in vol. xxiii. of the Med. Chir. Society's *Transactions*.

of his own case, when it is one of the cystin diathesis, I quote
the following:—" I am still (Nov. 1864) bothered with the
cystin, as you will readily understand when I tell you that
Syme(b) crushed a calculus for me about four months ago, and
removed another between two and three months since by the
median operation." The first of these calculi must have been
at least fifteen or sixteen years in forming.

As regards diet, he remarks :—" The only change I had made
in my diet during the last two or three years was to take more
wine than I ever did before. I generally took about three
glasses of dry sherry at dinner ; and although it helped me to
do more work, I am sure I was on the whole the worse for
it." . . . " Wine, beef, and mutton seemed to increase the
deposit; so that I have recently given up the use of all
stimulants, and not tasted beef or mutton above two or three
times during the last three months. Fish and vegetables I do
not consider do me any harm ; but the only vegetable I take
is potato, and that, too, but once a day. Fish, on the other
hand, I take every morning to breakfast, and at dinner I
usually limit myself to fowl, game, or veal. The great point
is not to eat more than the wants of the body demand. Both
overwork and mental worry increase cystin deposits."

XANTHIN
AS A URINARY DEPOSIT, AND AS CALCULI.

Xanthin was first discovered in a calculus weighing eight
grains by Marcet, senior, and since then, although occasionally
met with in the urine as a sediment, it has only on two other
occasions been found in the form of a stone in the human body.

Chemistry.

Xanthin, when pure, is a white, beeswax-like non-crystal-
line organic substance, insoluble in cold, but slightly soluble
in hot water. In 100 parts it consists of—

Carbon, 39·5 ; Nitrogen, 36 8 ; Oxygen, 21·1 ; Hydrogen, 2·6.

(b) The late Professor Syme, of Edinburgh.

It differs from cystin, therefore, in containing no sulphur, and from uric acid in its small proportion of nitrogen (uric acid contains over 33 per cent.).

Xanthin is soluble in ammonia, and very readily dissolved in caustic potash. From its alkaline solutions it is precipitated by acids, even by a current of carbonic acid. It is also soluble in the strong mineral acids with the aid of heat, and its solution in hydrochloric acid on cooling yields microscopic crystals of the hydrochlorate of xanthin.

On adding nitrate of silver to the nitric acid solution of xanthin, a crystalline, fine needle-shaped precipitate falls down.

When heated on a spatula with nitric acid, as in testing for uric acid, xanthin forms a yellow substance, which, on the addition of potash and evaporation, gives a dark purple-red colour.

Xanthin can easily be separated from uric acid by dissolving the whole in pure sulphuric acid, and then adding distilled water, which instantly precipitates the uric acid while it leaves the xanthin in solution.

Physiology.

Xanthin is pretty widely distributed throughout the animal organism. It has been found in very many healthy tissues of the human body. In the brain, liver, spleen, pancreas, thymus gland, the muscles, and the blood; and it is supposed to be one of the intermediate products of the metamorphosis of protein substances.

Xanthin is closely allied to guanin, which is also to be met with in the human body, and it can be prepared artificially from that substance by the aid of nitric acid.

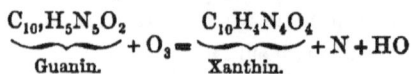

$$\underbrace{C_{10}H_5N_5O_2}_{\text{Guanin.}} + O_3 = \underbrace{C_{10}H_4N_4O_4}_{\text{Xanthin.}} + N + HO$$

Scherer says xanthin is a normal constituent of healthy human urine, but is not met with in sufficient quantity to render its detection sufficiently easy for clinical purposes. Neubauer thinks that on an average there is about 1 gramme (15½ grains) in 600 pounds of human urine. In the urine of spiders, on the other hand, it exists in abundance (Davy), and the concretions called Oriental bézoards also contain a large quantity of this substance.

Pathology.

Xanthin is occasionally met with in diseases as a urinary sediment, and more rarely still as a calculus.

Vogel speaks of the xanthin being much increased in cases of enlarged spleen, and also in acute atrophy of the liver. Douglas Maclagan(a) believes he found it in the urine of an hysterical girl ; and Jackson(b) thinks he met with it in a case of diabetes. Bence Jones(c) has lately described a case in which a crystalline deposit of xanthin repeatedly appeared in the urine. The case is briefly as follows :—

A schoolboy, aged 9½ years, enjoyed health up to the age of 7, when he was seized with violent sickness and pain in the stomach. On the third day the urine became of the colour of blood. It then stopped, and none was passed for four days, when delirium and convulsions came on. These lasted twenty-four hours, ceasing after a violent perspiration. No stone was known to have passed. In a few days he got perfectly well, and continued so for three years, when he caught a slight cold, and was delirious one night. The urine contained albumen, and was generally thick, and of high specific gravity. Under the microscope, the sediment was found to contain small oblong plates resembling uric acid (Fig. 20, a). But as these were dissolved by boiling, and soluble in hydrochloric acid, as well as in alkalies, and the hydrochloric acid solution,

---

(a) *Edin. Med. Journ.*, 1851.　　(b) *Arch. der Pharm.*, 1837.
(c) *Journ. of the Chem. Soc.*, 1862.

on being evaporated to dryness, yielded prismatic and hexagonal crystals (Fig. 20, *b*), the deposit was supposed to be
xanthin, with whose reactions, as is seen, it closely agrees.

Fig. 20.

As already said, xanthin
has been met with in the
form of calculi in the human
subject; but cases of the kind
are so exceedingly rare that
as yet only three have been
recorded.   This specimen,
consisting of a few fragments,
formed part of the celebrated
xanthin calculus, the size of
a small hen's egg, which was

FIG. 20.—Xanthin from human urine.
(*a*.) Spontaneous deposit.   (*b*.) Dissolved in hydrochloric acid and re-
crystallised.

successfully extracted by Professor Langenbeck from the
bladder of a Hanoverian peasant boy aged eight years.   The
stone broke in the process of extraction, and these portions
were presented by Langenbeck to our late Professor of Surgery,
Mr. Liston.   While analysing the calculi in our museum a few
years ago I had occasion to examine this specimen, and found
that it possessed the following characters : It is of a dark
cinnamon-red colour, exceedingly hard, of somewhat glistening
fracture, and distinctly laminated.

Small fragments are readily dissolved in caustic potash, less
so in ammonia, scarcely at all in water, even after prolonged
boiling, and very sparingly soluble in the mineral acids.   The
much slighter solubility of the calculus, when compared to
pure artificially prepared xanthin, arises no doubt from its
exceedingly dense nature, for when finely powdered its solubility is greatly increased.   On incinerating a small fragment
it slowly charred, without any flame, or disagreeable odour,
thus presenting a striking difference to cystin, and was at
length gradually dissipated without leaving any perceptible
inorganic residue.

The peculiar red colour of the calculus is no doubt entirely due to the xanthin having taken up a quantity of the colouring matter (urohæmatin) of the urine, just as uric acid under similar circumstances does ; for pure xanthin is white.

LEUCIN : ITS CLINICAL SIGNIFICANCE.

Chemistry.

Leucin is, when pure, a white non-crystallisable, odourless, and tasteless organic fatty-looking substance, composed in 100 parts of

Carbon	54·96
Oxygen	24·44
Nitrogen	10·68
Hydrogen	9·92
	100·00

Although looking something like fatty matter, it is quite different from it in its chemical reactions, for it is very soluble in water, sparingly soluble in alcohol, and quite insoluble in ether. Moreover, it is soluble in strong acids as well as in alkalies.

Leucin can be artificially prepared from almost any protein substance by decomposing it with acids, alkalies, or even by the process of fermentation. Leucin is itself also readily changed by oxidation, and when fully oxidised is entirely transformed into ammonia, water, and carbonic acid.

The best chemical test for leucin is that proposed by Scherer —namely, to put a small quantity on a platinum spatula, add nitric acid, evaporate to dryness, and then treat the residue with caustic soda, which dissolves it. When the solution thus obtained is concentrated, an oily-looking drop is formed, which can be readily rolled about on the spatula.

In the urine, leucin is never met with in sufficient quantity

to allow of the application of this test; we have therefore to
content ourselves with its microscopic appearances, which
are tolerably faithfully represented in the accompanying
woodcut.

FIG. 21.

FIG. 21.—Leucin from concentrated human urine in a case of chronic
atrophy of the liver.

The circular oily-looking discs might at first glance be mistaken for oil, not only on account of their microscopic characters,
but also on account of their being lighter than water. Their
insolubility in ether at once distinguishes them from all oily
matters. Besides this, although the discs show no crystalline
characters, they are occasionally laminated like the granules
of potato starch, and hence are apt to be confounded with
microscopic crystals of carbonate of lime. The lime crystals,
however, sink, whereas the greater number of globules of
leucin float, in water.

Leucin, as met with in urine, is always deeply impregnated
with the colouring matter of the liquid; hence, in cases of
jaundice, the globules have a dark yellow hue.

In searching for leucin in human urine, an ounce or two of
the liquid should be slowly evaporated to the consistence of a
syrup, set aside to cool, and afterwards examined with the
microscope. As yet we have not the means of making a quan-

titative analysis of this substance with sufficient facility for clinical purposes.

Physiology.

Leucin has for many years been known to scientific men as a normal product of some of the animal organs, but it is only within the last three or four years that it has assumed a clinical importance.

According to the researches of Liebig, Scherer, Gorup-Besanez, Frerichs, Neukomm, and many others, it may now be said to be one of the normal constituents of the spleen, liver, pancreas, lungs, brain, thymus, and thyroid; but whether it exists in the healthy living body as such, or is only a morbid product of decomposition, is not yet a settled question. Be that as it may, one thing is clear—namely, that its quantity is greatly increased during disease, and that its appearance in some of the excretions is a diagnostic sign of considerable value. Leucin is considered to be one of the transition products in the metamorphosis of albuminoid substances, and this, in the majority of instances, is no doubt true; but I believe it also to be, under certain circumstances, the direct product of the interrupted metamorphosis of hepatic substances into taurocholic acid. The reason for this opinion I shall presently briefly give, when speaking of the origin of tyrosin, a substance with which it is intimately associated.

Pathology.

Frerichs was the first to point out the diagnostic value of the presence of leucin in the urine in hepatic disease. He showed that the urine contained it in abundance in cases of acute atrophy, which observation has since been confirmed by Scherer, Neukomm, Wilks, myself, and some others. Moreover, I have recently shown that leucin is also to be detected in the urine in cases of chronic atrophy of the liver.

It is also occasionally to be met with in the urine in cases of typhus associated with jaundice (Frerichs, Murchison), and in a few severe cases of small-pox.

Leucin has been met with in abundance, in the liver in the above-mentioned cases, as well as in some cases of phthisis and rheumatic fever, and the natural conclusion is that the hepatic organ is the one which excretes it in greatest quantity. Indeed, at present, it is only in cases of hepatic disease that the existence of leucin in the renal secretion can be regarded as of diagnostic value. It then points out the existence of serious mischief, and is anything but a favourable sign for prognosis, all the cases in which it has been as yet detected having run on to a fatal termination. As regards treatment nothing need be said, as it is not the leucin deposit, but the disease inducing it we have to treat.

TYROSIN—ITS CLINICAL SIGNIFICANCE.
Chemistry.

Tyrosin belongs to the same class of substances as leucin, and like it may be artificially obtained from almost any protein matter. Its composition is,—

Carbon	59·67
Oxygen	26·52
Nitrogen	7·73
Hydrogen	6·08
	100·00

Tyrosin differs from leucin in being crystallisable. When pure it crystallises in fine white glistening stellate groups of small prisms, as represented in Fig. 22.

Tyrosin is odourless and tasteless, but when burned gives off a disagreeable smelling vapour. It is soluble in acids, alkalies, and boiling water; insoluble in cold water, ether, and alcohol.

The easiest applied chemical test for it is that proposed by

Scherer—namely, to moisten a few crystals with strong nitric acid on a spatula, and slowly evaporate to dryness. The residue, which is of a fine rich yellow colour, when moistened with hydrochloric acid becomes red. If the mixture be again evaporated to dryness, it yields a blackish residue. Reactions which are thought to be quite characteristic of tyrosin.

FIG. 22.

(a)

(b)   (c)

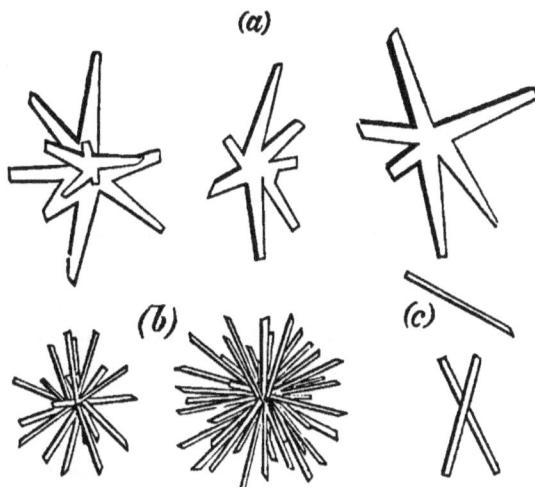

FIG. 22.—Crystals of pure tyrosin obtained from the urine of a case of chronic atrophy of the liver. (a) Large crystals; (b) the common form of stellate groups of needle-shaped crystals; (c) a few separate prisms (600 diam.).

As met with in urine, it is usually so mixed with leucin and other matters, that before applying the test it is necessary to purify it. This may be done in the following manner:—

Precipitate all the colouring matter by means of a solution of basic acetate of lead; filter, and then free the liquid from the excess of lead by means of a current of sulphuretted hydrogen; again filter, and evaporate the clear liquid to nearly dryness, when the tyrosin will crystallise in the white stellate groups already described.

In testing for impure tyrosin, add to the suspected solution a little nitrate of the protoxide of mercury (nearly neutral), which will throw down a red precipitate, and turn the supernatant liquid rose coloured,if tyrosin be present (Hofmann).

Another means by which tyrosin may be recognised is that proposed by Frerichs. The suspected substance is put into a watch-glass along with sulphuric acid, and after standing half an hour the mixture is diluted with water. It is next boiled and neutralised with carbonate of soda, filtered, and to the clear filtrate a few drops of perchloride of iron devoid of acid is added, when the presence of tyrosin is recognised by the formation of a dark purple colour.

It seldom happens that tyrosin is present in human urine in sufficient quantity to admit of the proper application of these tests, so that we are forced, in the majority of cases, to confine ourselves to its microscopic appearances, which, fortunately for us, are amply sufficient for our purpose. In doubtful cases it is well to concentrate the urine, as recommended in the case of leucin.

Physiology.

Tyrosin is met with in, or rather may be obtained from, almost all the organs in which leucin is to be found. It can scarcely, however, be said to be a normal constituent of the human frame, but rather appears to be one of the artificial products of the decomposition of highly nitrogenised matters. It may be artificially obtained in considerable quantities by acting upon horn, hair, or feathers, with sulphuric acid.

Pathology.

As a pathological product tyrosin is occasionally met with in the kidneys, in the urine, and in the liver in a free state. In acute yellow atrophy it may be almost said to be constantly present in all of these situations, for Staedeler,

Frerichs, Neukomm, Scherer, myself, and several others have met with well-formed crystals of tyrosin in the tissue of the atrophied liver, of the kidneys, and in the urine, without subjecting either the one or the other to any chemical manipulation, but merely examining them with the microscope.

The crystals in these cases do not always assume the shape represented at Fig. 22, but occasionally form very thickly-set stellate groups of fine needles, or spiculated balls, not unlike a rolled-up hedgehog with the bristles sticking out in all directions, as represented in the accompanying woodcut.

FIG. 23.

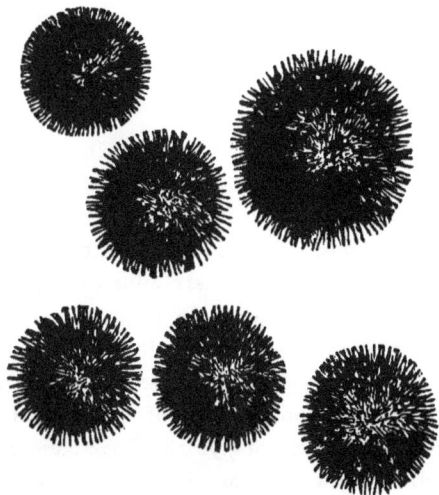

FIG. 23.—Spiculated balls of tyrosin, from the urine of a case of acute atrophy of the liver. When these were re-dissolved, purified, and re-crystallised, they assumed the form of those in Fig. 22 (mag. 600 diam.).

The crystals found in the urine are usually of a deep yellow colour, in consequence of their having taken up a quantity of the bile pigment. Tyrosin has also been detected in the urine in cases of typhus complicated with jaundice, and in severe cases of small-pox, as well as in at least one example of chronic hepatic atrophy supervening on obstructed gall-duct. Very recently Murchison and myself found free crystals of tyrosin in the tissue of the kidneys of a private soldier,. who died comatose, and whose liver was found to be much below the ordinary weight.

The last question to be considered is : "Where does the tyrosin met with in disease come from ?" The results of my experiments on animals, to which jaundice had been artificially given, and which have been elsewhere published, led me to the belief that tyrosin and leucin stand in the same relation to each other in disease as the two bile acids, glycocholic and taurocholic, do in health. Glycocholic acid is crystallisable, taurocholic is not. '

The urine of some of the dogs to which I gave artificial jaundice by the subcutaneous injection of bile contained both leucin and tyrosin, and even in one of these cases crystals of tyrosin spontaneously formed in the bile taken from the animal's gall-bladder immediately after death, and merely allowed slowly to evaporate. In another case, again, free crystals of tyrosin were encountered in the tissue of the liver itself, all of which facts led me to the conclusion that tyrosin and leucin are the products either of the arrested or of the retrograde metamorphosis of glycocholic and taurocholic acids.

Be the theory, however, what it may, the important fact for us clinical Physicians to bear in mind is, that the presence of tyrosin in the urine of any of our patients is an almost certain sign of a rapidly approaching fatal termination.

LECTURE X.

DIABETES MELLITUS—GLUCOSURIA—SACCHARINE URINE—ITS
NATURE AND RATIONAL TREATMENT.

In no department of experimental research have results of
greater value to the clinical Physician been recently obtained
than in that connected with the study of animal saccharine
matter. Within the last few years, indeed (since 1848), an
entire revolution has taken place in our ideas of the physiology
and pathology of diabetes ; and although we have still much
to learn before we can completely unravel the skein of laws
regulating the formation and destruction of sugar in the animal
economy, we have, nevertheless, reason to congratulate ourselves upon that which has already been achieved, and be
sanguine in our expectations of what science may yet
accomplish.

Before entering upon the consideration of this important
era, which dawned within my own time, and whose gradual
development I have not only watched as an interested spec-
tator, but occasionally assisted as a conscientious actor, it will
be advisable for me briefly to scan the earlier literature of
diabetes, in order that you may better appreciate the advances
that have been recently made, and comprehend in what respect
the views of its pathology about to be given essentially differ
from those of my predecessors and contemporaries.

From time immemorial, cases of emaciation, accompanied
by an inordinate thirst and voracious appetite, had been

observed, and in consequence of the patients so affected being at the same time troubled with an excessive elimination of urine, ancient Physicians gave to the disease the name of Diabetes (δια, through ; βαινω, I go). It was not, however, until 1674 that the urine was discovered to possess, in some cases, a sweet taste, the honour of which discovery belongs to an English Physician named Thomas Willis. From this time henceforth the disease was divided into two classes, one of which received the name of Diabetes insipidus (without sugar), the other that of Diabetes mellitus (with sugar).

In 1774, exactly one hundred years after the date just alluded to, Mathew Dobson, a Physician practising in Liverpool, discovered that the blood as well as the urine in diabetes contains sugar, and from this observation he justly concluded that the saccharine matter found in the urine is not formed in, but only excreted by, the kidneys.

In 1778, Cowley succeeded in separating the sugar from the urine in a free state. (I may here remark that Bartholdi, so early as 1619, called attention to the presence of saccharine matter in milk ; but this is, of course, a point entirely unconnected with diabetes.)

In 1796, John Rollo, Surgeon-General to the Royal Artillery, made the first important observation regarding the treatment of diabetes by discovering that an animal diet not only reduces the quantity of urine, but even diminishes the amount of sugar daily eliminated.

The next two steps were made by foreigners.

In 1815, M. Chevreul ascertained that the saccharine matter met with in diabetic urine differs from ordinary cane sugar, and closely resembles that of the grape.

In 1825, another important step was gained by Tiedemann and Gmelin discovering that starch is transformed into sugar during its passage along the alimentary canal.

In 1837, the next observation of interest was made by

M'Gregor, of Glasgow, who found sugar in the vomited matters of diabetic patients—an observation which seemed to confirm Rollo's idea that the disease arises from the gastric juice turning vegetable food into sugar; and from that time till the present animal diet was consequently considered our sheet anchor.

We now arrive at an entirely new phase in the literature of diabetes, in which the teachings of the sick chamber gave place to those of the laboratory.

In 1848 the physiological world seemed as if struck by a thunderbolt when Bernard proclaimed that animals, as well as vegetables, had a sugar-creating power. Until then all the saccharine matter met with in the human body, whether in health or in disease, was supposed to originate in the transformation of vegetable substances. And now, for the first time, were we made alive to the startling fact that men, like sugar-canes, possess within themselves a saccharine manufactory; the liver being daily and hourly as actively engaged in fabricating sugar as in secreting bile.

Like all great discoveries, the rays of this one were not limited to its own field of inquiry, but broadly reflected over the various departments of scientific Medicine. By it new ideas were awakened, new discoveries made, and even at the present hour the impetus it gave to original research is still alike perceptible in the laboratory, in the dead-house, and in the clinical ward. In connexion with the subject of diabetes alone, the following may be reckoned as a few of the more important discoveries to which it led :—

In 1849 Bernard discovered that the disease can be artificially communicated to animals by pricking the floor of the fourth cerebral ventricle.

In 1853 I discovered that diabetes may be artificially induced in animals by exciting the liver through means of stimulants, such as alcohol, directly introduced into the portal circulation.

An observation which explains the well-known fact that
diabetes is a much more common disease in spirit-drinking
than other countries.

In 1855 Bernard discovered that the formation of sugar in
the liver cannot be regarded in the light of a "vital" process,
as it goes on, not only after the death of the animal, but even
after the removal of the liver from the body.

In 1856-7, Chauveau and myself gained another piece of
ground by ascertaining that the sugar normally present in the
circulation is not burned off in the lungs, as hitherto supposed,
but disappears from the blood in its transit through the capillaries of the general circulation. The function of the saccharine matter most probably being to nourish the body.

In 1857, Bernard made the additional discovery, that before
albuminous substances are converted into sugar, they first
pass through the transitional stage of glucogene (animal starch).

Lastly, in 1859-60, Brücke and Jones ascertained, by careful
experiment, that traces of sugar are even to be detected
in normal urine—an observation which, we shall afterwards
see, has an important bearing on the pathology of diabetes ;
for, as I stated in a previous lecture, it may be regarded as a
fundamental law that in disease neither new substances nor
new functions are created. Morbid phenomena being merely
the result of a change in the quantity and quality of normally
existing agents and agencies.

Chemistry.

Although every one knows, or at least imagines he knows,
a good deal about the nature and properties of sugar, I shall
venture to say a few words upon the subject ; for the more
extensive our knowledge, the easier will be the comprehension
of this hitherto-regarded inexplicable disease. And on the
present occasion it is the more necessary for me to do so,

seeing that I differ from my predecessors and contemporaries in believing that there are at least two distinct forms of the disease, requiring diametrically opposite lines of treatment— an opinion which I shall presently show is as much in accordance with the teachings of the bedside as of those of the laboratory.

The term "sugar" is applied to a great variety of substances, the common essential character being their sweet taste. Chemists have, however, drawn broad distinctions between the different forms of saccharine matters, the properties of which are exceedingly varied. But for the sake of convenience they have been divided into two great classes—a *first* and a *second*—according to the manner in which they are acted upon by acids, and alkalies.

To the first class, of which *cane sugar* is the type, are given all those that are easily crystallised, and which when boiled with an alkali are not decomposed, but enter into combination with it to form a saccharate of the alkali.

To the second class, of which *grape sugar* is the type, are awarded those which are not easily crystallised, and which when boiled with an alkali are transformed into the acids of molasses—glucic and melassic acids.

When acted upon by acids, on the other hand, an equally remarkable feature of distinction is observable in these two classes of sugar; for while sugars of the second class remain apparently unaffected, sugars of the first (cane) are rapidly transformed into those of the second class (grape).

These changes are easily illustrated. If, for example, I boil a little sugar candy (a sugar of the first class) with potash, no apparent change is visible; whereas, if before boiling it with potash I heat the solution of sugar candy in a test tube along with a couple of drops of hydrochloric acid, on adding the potash, the whole liquid becomes of a dark brown colour in consequence of the grape sugar (a sugar of the second class) which was formed by the action of the hydrochloric acid upon the cane sugar being instantly decomposed by the potash into the acids of molasses (glucic and melassic).

Just as the vegetable kingdom furnishes us with the two great types of sugar, *cane* and *grape*, so the animal kingdom yields to us their representatives in the form of milk, and liver sugar ; milk sugar being the analogue of cane, liver sugar of grape. For example, the sugar of milk is readily crystallised, is unaffected by alkalies, but easily transformed into liver sugar by the action of acids. Liver sugar, on the other hand, is crystallised with difficulty, unaffected by acids, but rapidly decomposed by alkalies. (a)

Even the relative distribution of the two classes of sugar in the vegetable and animal kingdom is nearly identical ; for just as in the vegetable world the sugars of the second greatly preponderate over those of the first class, so in the animal economy, while the second class sugar is to be encountered in the blood, the liver, and the urine, that of the first class is limited to the mammary secretion alone.

Tests for Saccharine Urine.—A great variety have been proposed, but I shall only enumerate those possessing a real practical value.

1st. Specific Gravity.—Saccharine urine is generally, though not always, of a high specific gravity, averaging from 1020 to 1040. The specific gravity may be much higher, or much lower, even independently of the amount of sugar it contains. For in diabetic urine there is generally a large excess of urea,

----

(a) By combining liver sugar with chloride of sodium (common salt), Berthelot and De Luca have succeeded in obtaining large, colourless, transparent crystals, the watery solution of which ferments with yeast, and reduces the oxide of copper. The composition of the crystals, as ascertained by analysis, is represented by the accompanying formula— $2C_{12}H_{12}O_{12}, 2HO + NaCl$. The authors conclude therefrom that liver sugar is not only exactly the same as that found in diabetic urine, but that it is also identical with the sugar of the grape.—*Compt.-Rend.*, vol. xlix., p. 213.

which greatly adds to its specific gravity; and occasionally, on the other hand, the disease is accompanied with an albuminuria sufficient to reduce the specific gravity even to below the normal standard. The specific gravity itself is therefore no criterion of the amount of sugar contained in any given specimen of urine.

2nd. Potash Test.—To a drachm of urine in a test-tube add an equal bulk of a solution of potash of the specific gravity of 1060. Heat the upper half of the liquid in the flame of the lamp until it boils; then mark the change. If sugar be present, the upper boiled portion of the liquid changes colour, and becomes yellow or brown, in proportion to the amount of saccharine matter in the liquid. The browning being due to the transformation of the sugar into glucic and melassic acids. The lower cold half of the liquid, on the other hand, by remaining unaffected serves as a standard of comparison.

3rd. Copper Test.—To a drachm of urine add half a drachm of the potash solution. Shake the mixture and *afterwards* add a few drops of a solution of sulphate of copper (a); just sufficient to produce a pale blue tint when the whole is agitated. Boil the liquid from the bottom, and if sugar be present the blue colour will disappear, and a yellow or red precipitate form, according to the amount of sugar present in the urine.

The chemical steps of this process are as follows:—When sulphate of copper (SO_3CuO) is added to a solution of caustic potash (KaO), the sulphuric acid combines with the potash to form a soluble sulphate, and the oxide of copper separates in the form of a gelatinous precipitate, which, however, in consequence of the presence of organic matter remains suspended in the urine.

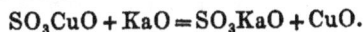

$$SO_3CuO + KaO = SO_3KaO + CuO.$$

(a) The most convenient strength I find is ten grains to the ounce of water.

The next stage in the process is produced by the heat causing the sugar to oxidise itself at the expense of the oxide of copper; whereby the yellow suboxide is thrown down.

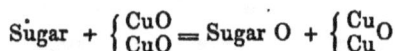

$$\text{Sugar} + \begin{cases} CuO \\ CuO \end{cases} = \text{Sugar } O + \begin{cases} Cu \\ Cu \end{cases} O$$

If much sugar be present, and the boiling is continued sufficiently long, all the oxygen is withdrawn from the copper, and the red metal is precipitated on the bottom and sides cf the test-tube.

N.B.—Should the urine contain albumen, the albumen must be separated by the aid of heat and acetic acid before applying the copper test. Otherwise, on adding the sulphate of copper to the urine + the potash, a mauve instead of a blue liquid will be obtained; and, on boiling, no reduction of the copper will occur unless a large excess of sugar be present, the liquid merely changing its tint from a mauve to a purple, and from a purple to a red colour.

When the urine contains bile pigment, or is otherwise high coloured, and the sugar is present in small quantity only, it is necessary to decolourise the urine before adding the reagents. For this purpose, put an ounce or two of urine into a six-ounce bottle along with a tablespoonful of animal charcoal and a small pinch of carbonate of soda. Shake the mixture well for a few minutes, and then filter. A perfectly colourless liquid will thus be obtained, and greatly facilitate the application of both the copper and potash tests.

A great deal has been said about the fallacies likely to arise from the presence of uric acid, cotton fibres, chloroform, etc., in the urine. All I can say is, that during a fifteen years' experience I have never encountered them. In all cases of doubt, however, in addition to the potash and copper, I invariably apply the fermentation test.

4th. Fermentation Test. — The easiest way of applying

this to urine is to employ an apparatus like that represented
in the accompanying woodcut.

FIG. 24.

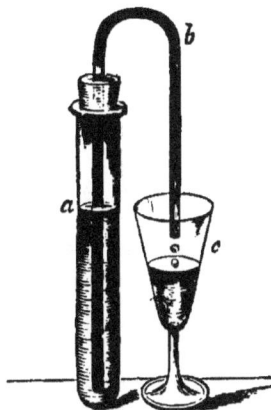

A test-tube, of about six inches
long and three-quarters of an inch
in diameter, is to be fitted by means
of a piece of cork with a piece of ordinary small glass tubing bent to the
shape of a syphon in the flame of the
spirit-lamp. When ready for use,
the tube is filled brimfull of urine,
to which a few drops of baker's
yeast have been added. The syphon
is next fixed in the tube by means
of the well-fitting cork, and the
free end allowed to dip into a glass
conveniently placed to receive the
liquid as it is driven out of the tube by the carbonic acid
generated during the process of fermentation. Each equivalent
of sugar is transformed into one of alcohol and four of carbonic
acid gas, so that every seventeen cubic centimetres of carbonic
acid evolved is equivalent to one grain of sugar decomposed.

During the fermentation process the apparatus must be
kept in a moderately warm room.

When diabetic urine is set aside for a few days it frequently
ferments spontaneously, and the liquid becomes filled with
torulæ, or sugar fungi, which can be readily recognised with
the microscope. They consist, like the common yeast plant,
of a number of spores strung together in short rows like beads.
The sporules of torulæ are, however, smaller than those of
yeast, their relative size being represented in Fig. 25. In a
few cases I have seen spores of torulæ spontaneously form in
urine within twenty-four hours after its emission, although,
as a rule, about thirty-six hours are necessary for this
development.

FIG. 25.

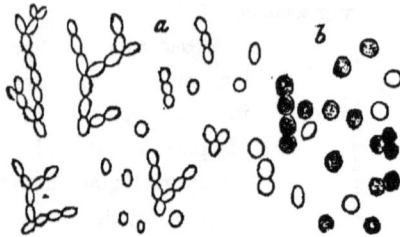

FIG. 25.—a. Sporules of Torula Cervisiæ from diabetic urine. b. Spores from baker's yeast.

Quantitative Analysis.—At one time the quantitative determination of sugar was a long and tedious process, but now by adopting the following method I find it both quick and easy. Three solutions must be retained in readiness :—

1st. A standard solution of sulphate of copper, made by dissolving 34·63 grammes (536·76 grains) of dry sulphate of copper in 1000 c. c. (32·26 oz.) of distilled water. 1 c. c. of this solution represents 0·005 gramme of dry diabetic sugar.

2nd. A solution of caustic potash of the specific gravity of 1060.

3rd. A saturated solution of the bitartrate of potash.

The object of keeping these solutions separate, instead of combining them as is usually done, is in order to avoid the danger of the spontaneous reduction of the oxide of copper, which invariably takes place sooner or later in all the ordinary standard solutions. Having the solutions at hand, place before you two porcelain capsules, each containing an ounce of distilled water ; to which add 20 c. c. (measured in glass A, Fig. 8) of each of the solutions, first adding the copper, then the bitartrate, and lastly the caustic potash. Next take the measure B, Fig. 4, and pour into it 10 c. c. of urine, and dilute with 90 c. c. of water. The mixture will thus amount to 100 c. c.

The succeeding step is to put one of the capsules over the

spirit lamp, and as soon as its contents begin to boil add, drop by drop, the diluted urine until the blue colour of the liquid entirely disappears. Then place it aside in order that the reduced copper may fall to the bottom, and allow you to judge by the depth of colour of the supernatant liquid if much of the oxide of copper still remains to be precipitated. While this is going on, boil the contents of the second capsule after adding to it an amount of dilute urine equal to that which was employed in the first experiment; and should the supernatant liquid in the first capsule turn out to be still blue, add, in addition to the amount already used, 5 c. c. of dilute urine. After boiling for a few minutes, place it aside in order that its reduced copper may fall down, and repeat the same thing with the other capsule until the exact quantity of dilute urine required to reduce the 20 c. c. of standard copper solution be ascertained. So soon as that is accomplished, all that is required in order to find out the quantity of sugar eliminated by the patient in twenty-four hours is to divide the total number of c. c. of urine passed by the number of c. c. of dilute urine required to reduce the 20 c. c. of copper solution.

Suppose, for example, the patient passed, as in case now before me, 3720 c. c. of urine in twenty-four hours, and 26 c. c. of dilute urine were required to reduce the oxide of copper, the 3720 divided by 26 would give the number of grammes of sugar contained in the twenty-four hours' urine.

$3720 \div 26 = 143$ grammes in the 24 hours' urine.

Should the urine contain very little sugar, it is quite unnecessary to dilute it; and in that case, the calculation is made in the usual way, as for example, for urea.

Those who have not the means of making a quantitative analysis in the way above described may adopt the simpler, though less exact, method proposed by Dr. William Roberts. The following is the mode of procedure :—

1. Four ounces of urine are placed in a twelve-ounce phial, with a lump of German yeast of the size of a small walnut.

2. This is loosely corked, or covered with a slip of glass, and placed in a warm place to ferment.

3. A companion phial filled with the same urine—say a three-ounce phial—is tightly corked, and placed beside the fermenting phial.

4. In about twenty-two hours, when fermentation has ceased, the two phials are removed, and placed in some cooler part of the room.

5. Two hours after—that is, about twenty-four hours from the commencement of the experiment—the contents of the phials are separately poured into cylindrical glasses, and the density of each observed.

6. The difference between the two specific gravities is thus ascertained, and every degree of " density lost " indicates one grain of sugar per fluid ounce of the urine.

Crystals of diabetic sugar may be prepared by simply evaporating a few drops of urine to dryness on a glass slide. But this is only possible when the urine is very rich in sugar, and contains but little urea, and other salts. The most characteristic form of crystal is that of the rhomboidal prism occasionally arranged in arborescent tufts, as represented in the accompanying drawing (Fig. 26), which was kindly made for me by Mr. T. R. Loy, one of the gentlemen attending my practical class. Such crystals are in general called diabetic sugar ; but I believe them to be a combination of sugar and chloride of sodium, for except in the presence of chloride of sodium diabetic sugar never assumes a so regularly prismatic form of crystallisation. Gibb says that when diabetic urine contains a larger proportion of salts, the sugar crystallises in little circular masses, with minute crystals projecting from the surface. The masses appear to be made up of an aggregation of flat plates of sugar, and, when examined on a dark ground, resemble lumps of the well-known barley-sugar.

FIG. 26.

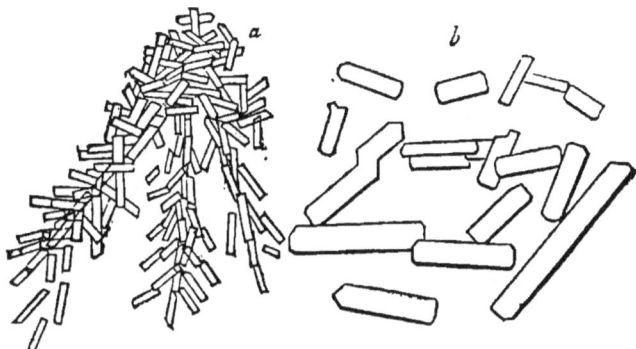

FIG. 26.—Compound crystals of diabetic sugar and common salt, spontaneously formed in concentrated human urine. *a* As seen under a low power. *b*. Highly magnified.

Physiology.

If we desire to be philosophical as well as practical, it is necessary that we should thoroughly understand the origin and destruction of the sugar met with in the animal economy. I shall, therefore, now proceed to the consideration of the *physiology* of diabetes.

That sugar is a normal constituent of the human frame is easily shown by withdrawing an ounce of blood from a healthy man in full digestion, and allowing it to fall drop by drop into a couple of ounces of boiling water faintly acidulated with acetic acid. By so doing all the albuminous matters are so firmly coagulated, that on filtration a perfectly colourless liquid is obtained, and on applying to it the copper, potash, and fermentation tests, the existence of sugar can be demonstrated with facility. Should the patient have had a mixed meal, a great part of the sugar found in his blood will have been directly obtained from the food. This opinion is opposed to the view of Bernard, who on finding that animals have the power of making sugar out of albuminous substances, fell into the error of supposing that all the saccharine matter met with in the human body must of necessity be the product of the liver. The results of my experiments on animals have led me to an entirely opposite conclusion in as far as the omnivora and herbivora are concerned. This opinion I arrived at from finding abundance of saccharine matter in the blood of the portal vein and chyle of the thoracic duct of a dog, three hours after he had partaken of a quantity of horse-flesh, to which had been added a couple of ounces of soft sugar.

Even in those cases where the patient has had a mixed diet, without, however, having partaken of any sugar in a free state, I still maintain that a great part of the saccharine matter met with in the blood is the direct product of the food, for the following reason :—All vegetable foods, such as bread, turnips, carrots, potatoes, etc., contain a large quantity of starch, which starch during its passage along the alimentary canal is all, or nearly all, converted into sugar through the agency of the digestive juices, especially of the saliva and pancreatic fluids.

The transformation of amylaceous matters into glucose during their sojourn in the alimentary canal is not the result of accident, but the sequence of an unalterable law, which is equally in force in or out of the body, so long as the physical conditions necessary for its action are in operation. Thus, if an ounce of boiled arrowroot, in a test-tube, be mixed with a little saliva, or have added to it half a dozen drops of pancreatic juice, all the starch contained in the arrowroot will infallibly be changed into sugar in the course of a few minutes. And for this operation the intervention of no other agency than that of a temperature equal to that of the human body is required. A precisely similar change occurs in the animal organism, and the sugar thus formed is absorbed by the mesenteric veins and lacteals, and of necessity forms part of that met with in the circulation.

As regards carnivora, on the other hand, Bernard's assertion is strictly correct, as the result of the following experiment will prove. The experiment was most carefully performed some years ago by Professor Sharpey and myself, with the view of ascertaining not only if sugar exists in the blood of carnivora, but if sugar is actually present in the blood, of the healthy animal, at the moment of its withdrawal from the circulation.

From the femoral artery of a dog fed solely, during four days, on boiled flesh, perfectly devoid of sugar, one and a-half ounces of blood were allowed to flow directly into boiling water, acidulated with acetic acid, and when the clear filtrate from this blood was tested, it gave unmistakable evidence of sugar, which sugar must have been formed in the animal's body, seeing that not a particle of saccharine matter was introduced with the food. It is quite unnecessary for me to cite another experiment in order to prove that the animal organism has a sugar-creating power, for Nature has herself supplied the proofs. Is sugar not a constant ingredient of the normal milk of the flesh-eater, as well as of the vegetable feeder? Most assuredly it is. If the animal body cannot form sugar, from whence does the milk of the carnivora receive its supply? For example, where does the sugar found in the milk of the polar bear come from if it is not manufactured by some organ or other in the animal's own body? Not only does the polar bear live on animal diet, but on the flesh of animals, such as the walrus and seal, whose chief food is fish, which fish, in their turn, are not usually supposed to live upon vegetable matter. As regards this point, I am even inclined to go a step further than Bernard, and not only assert that the carnivorous animal has the power of forming sugar out of albuminous substances, but that even in the case of the herbivorous animal the sugar met with in its milk is not directly obtained from the digestive canal. This, I think, is proven by two facts.

1st. Milk sugar possesses certain special characters which distinguish it from all vegetable sugars.

2nd. Milk sugar, although abundantly present in milk, has not yet been detected in the circulation. The natural conclusion, therefore, is, that it is formed by the mammary gland.

Consequently, it is perfectly clear that there must be at least one organ in the body capable of forming sugar.

Now comes the question—In what organ of the body is the sugar formed which is met with in the general circulation of the carnivora? Answer—The liver. Why the liver? Because, in the case of the carnivora, the blood proceeding to the liver is devoid of sugar, while that coming from it is rich in saccharine matter—richer, indeed, than the blood of any other organ of the body. As Bernard may be considered in the light of an interested party, and I have already cited some of my own experiments, I shall now quote three from an entirely independent observer—Schmidt, of Dorpat; (a) the results of which are as follows :—

	Percentage of sugar in	
	Portal vein.	Hepatic vein.
1st dog (on animal diet) . .	0·00	0·93
2nd „ „ . .	0·00	0·99
3rd „ (fasting during two days) .	0·00	0·51

These results I have again and again confirmed, so that there is not the slightest doubt left in my mind regarding their validity, and I believe that they furnish us with the key to the well-known fact that some diabetic patients, even when

(a) *Compt. Rend.*, vol. xlix., p. 63.

totally restricted to animal diet, still pass a large quantity of sugar.

Having now ascertained that the healthy blood always contains sugar, and that when the saccharine matter is not obtained from without, the body manufactures it for itself, the next point is to determine if sugar be absolutely essential to life, and what use or uses it is put to in the animal organism.

We must begin by finding out if the quantity of sugar in the blood is always the same, or if it be liable to variation; and to what extent.

From the results of a series of experiments made many years ago, I came to the conclusion that the amount of sugar present in the arterial blood of healthy animals is subject to great fluctuations, varying from an almost inappreciable quantity, after long fasting, up to 0·24 per cent. during the time of full digestion. In the hepatic vein the amount, Bernard says, may even be as much as 2 per cent.

The quantity of sugar present in the general circulation seems to follow a definite law, for it goes on gradually increasing as digestion advances, and as gradually diminishing as we approach the period for the next meal,—the maximum being reached four or five hours after food, the minimum during fasting. Saccharine matter does not entirely disappear, however, from the circulation till after prolonged fasting. Chauveau found as much as 0·05 per cent. in the blood of a dog, and 0·09 per cent. in that of a horse kept during three entire days without food. In these cases, as the animals were of course forced to live on their own tissues, the sugar formed by their livers must have been made out of some one or other of the constituents of the blood.

The rise and fall in the amount of sugar in the circulation, according to the state of the digestion, is clearly the key to another often-observed pathological fact—namely, that in all cases of diabetes the amount of sugar in the urine is subject to great fluctuations during the course of the day; while it furthermore explains why in some slight cases of diabetes sugar is only to be found in the urine a few hours after a meal.

Having said that the amount of sugar in the blood fluctuates according to the stage of the digestion, it may now be added that the amount of saccharine matter present in the circulation varies in proportion to the quality and quantity of the food. The results of a series of experiments performed in 1858 before the Practical Physiology Class at University College furnish us with the following data :—

In the arterial blood of the carotid artery of three dogs, four hours after having been exclusively fed on horse-flesh, an average of 0·08 per cent. of sugar was found, whereas eight days subsequently, after these same dogs had been freely supplied with mixed diet, consisting of bread, meat, and potatoes, the average amount of sugar in the arterial circulation four hours after feeding rose to 0·22 per cent.,—hereby affording a physiological explanation of the well-known pathological fact that purely animal diet diminishes, while vegetable or mixed diet increases, the amount of saccharine matter eliminated by the kidneys.

I must now call attention to the fact that, although the quantity of sugar present in the circulation is constantly fluctuating, the amount found in the liver remains comparatively unaltered. Thus, for example, Bernard found, about the same period after death, an average of $1\frac{1}{2}$ per cent. of sugar in the livers of dogs fed exclusively on animal diet, on mixed food, and on vegetable matter,—a result which, from showing that the amount of sugar present in the liver is but slightly influenced by the kind of food, led him into the erroneous idea hat diet has no influence on the amount of saccharine matter manufactured by that organ ;—a mistake which has constantly brought him into difficulties while attempting to reconcile it with other entirely opposed facts,—such, for example, as the effects of different kinds of foods upon diabetic patients.

The reason why the amount of saccharine matter remains comparatively stationary in the liver while it is constantly fluctuating in the circulation arises from the circumstance of the sugar not being stored up in the hepatic cells, but poured into the vessels as quickly as it is formed; therefore the quantity which the liver at any time contains cannot be accepted as a criterion of the amount which the organ produces. Had Bernard gauged the amount of saccharine matter present in the blood, instead of in the liver, after each particular kind of diet, and drawn his conclusions from the results thereby obtained, he could not possibly have fallen into the above-mentioned error.

Having thus far traced the *rationale* of saccharine urine, we have still to advance a step further, and inquire whether animal food is directly transformed into sugar, or if there is a transition stage in the process between protein substances on the one hand, and saccharine matter on the other. In the case of amylaceous foods, we have already seen there is no connecting ink, but that, during their passage along the alimentary canal,' they are at once transformed into sugar by certain of the digestive juices. With protein substances, however, the case is very different; for, ere they can be changed into saccharine matter, they must first be resolved into something resembling starch.

I am constantly telling my students that Nature does nothing on a small scale; and the more we study her the more we are forced to admire the uniformity and extensive applicability of her laws. In the vegetable and animal kingdoms we not only encounter analogous substances, but we find that they are endowed with similar properties, and respectively perform similar functions. We already saw that while the sugar of the cane has its analogue in that of the milk, the sugar of the grape has its analogue in that of the liver. I have now to point out how even vegetable starch has also its analogue in the animal world.

Although the discovery of a substance in the animal body possessing the properties of starch had been long suspected, it was not until 1857 that a large enough quantity was separated in a sufficiently pure state to admit of its being subjected to elementary analysis, and thereby allow its striking affinities to vegetable starch to be demonstrated. To Bernard belongs the credit of having·first obtained this substance from the liver in a free state, and shown its direct convertibility into saccharine matter, in consequence of which he appropriately gave to it the title of glucogene.

Glucogene has such a striking resemblance to vegetable starch that a short description of its appearance and properties will not be out of place. Like ordinary starch it is a neutral, white, odourless, insipid substance ; soluble in water, but insoluble in alcohol and acetic acid. When boiled with diluted mineral acids, or treated with either saliva or pancreatic juice, it is transformed like other amyloid substances into sugar of the second class. That is to say, into a sugar possessing the property of reducing the oxide of copper, and of being decomposed by caustic potash into glucic and melassic acids. Moreover, it is also, like starch, transformed by fuming nitric acid into xyloidine, a combustible substance, which detonates with flame when heated to 180°.—Poggiale.(a) On the other hand, it differs from ordinary starch in giving with the tincture of iodine a reddish violet, instead of a deep blue colour. In which respect, therefore, it appears to stand between starch and dextrine.

The easiest mode of extracting glucogene from the liver is to add to a concentrated aqueous decoction of the hepatic

(a) Brown-Séquard's *Journal*, 1858, p. 549.

substance an excess of glacial acetic acid, which instantly precipitates all the glucogene, and leaves the albumen in solution. The precipitate is then separated by filtration, re-dissolved in water, and again precipitated by absolute alcohol; the process being repeated until the substance separates in a perfectly pure state.

While giving credit to Bernard for being the first to extract glucogene from the liver, I must not omit to yield to Hensen his proper meed of praise in having foreshadowed the discovery by his researches on the saccharine function of the liver published in 1856,(a) in which he stated that he had found in the liver of rabbits a yellow substance transformable into sugar by the action of saliva, pancreatic juice, and even portal blood.

According to Pelouze, glucogene is composed in 100 parts of

Oxygen	. 54·1
Carbon.	. 39·8
Hydrogen	. 6·1
	100·0

Glucogene has been detected in the livers of all animals in which it has hitherto been sought for, no matter whether they were animal or vegetable feeders, and consequently the natural conclusion is that the liver possesses the power of forming it out of animal as well as vegetable food. But while, in the case of carnivora, glucogene is only met with in the liver, in that of the herbivora it has been encountered in various other situations, for example, Poiseuille et Lefort found in the blood of a bull 0·073 per cent. In the chyle 0·123 per cent., and in the lymph 0·266 per cent.(b) It is even to be met with in the muscles. Here is a beautifully pure specimen of glucogene, given to me by Professor Scherer, which he obtained from the muscles of the horse. Moreover, the quantity of glucogene

(a) *Verh. Med. Phy. Gesell.*, Würzburg, vol. vii., part 11, 1856.
(b) *Comptes-Rendus*, April 5, 1858, p. 645.

in the livers of herbivora is many times greater than that met with in those of the carnivora.

Now comes the question, What are the uses of glucogene in the animal economy?

We all know that starch, as starch, cannot nourish the body, and from certain facts presently to be related it is highly probable that glucogene must, like ordinary starch, be transformed into saccharine matter before it can play its part in the intricate processes of life. We already know that at least a great part of the glucogene formed in the liver is changed into sugar before it is permitted to leave that organ. We are also acquainted with another exceedingly interesting and important fact, namely, that, be the transforming agency what it may, it is quite independent of any so-called vital influences. The transformation of glucogene into sugar appears to follow the same unalterable law as that already noticed regarding the transformation of ordinary starch, for not only does it continue in force after the death of the animal, but even after the removal of the liver from the body. All that is required apparently being the sustaining the organ at a certain temperature.

This assertion can be easily proved by taking the liver of a healthy animal, and after it has been, as far as possible, completely freed from sugar by passing a current of water through its vessels, keeping it in a warm apartment for six or eight hours. The tissues of the organ will then be found saturated with sugar, although only a small quantity could be detected immediately after its removal from the body.

The results of the following experiment,(a) which was performed by Professor Sharpey and myself, will give a tolerable idea of the rapidity with which glucogene is transformed into sugar in a liver after its removal from the body :—

(a) *Proceedings* of the Royal Society, No. 38, p. 289. *Lancet*, October 20, 1860, p. 336.

A dog which had been previously fed on animal diet received a full meal of bread and milk. Five hours afterwards the animal was pithed, and a portion of the liver rapidly sliced off and immersed in a freezing mixture. The portion of frozen liver was found to contain only 0·333 per cent. of saccharine matter; whereas another portion of the same liver, which had not been frozen, after having stood two hours was found to contain 1·55 per cent.—that is to say, nearly fivefold more of sugar, which sugar had, of course, formed at the expense of the glucogene after the death of the animal. The 0·333 per cent. represents, of course, the amount of sugar present in the liver at the moment of death—an amount which appears insignificant until it is recollected that during life the sugar formed in the liver is removed from it with every pulsation of the heart; consequently, the quantity poured into the circulation during the twenty-four hours must be very considerable, notwithstanding that during life only a fractional part of a per cent. is present in the organ at any one time.

I may here mention that it is this circumstance now alluded to, of there being but a very small quantity of sugar in the hepatic organ at the moment of death, while a large amount is to be found some hours afterwards, which has led Dr. Pavy into the error of supposing that the formation of sugar in the liver is in all cases due to a post-, instead of an ante-mortem cause. Were we to adopt such a mode of reasoning we might with equal justice say that no starch is changed into sugar in the digestive canal during life, for as its transformation likewise still continues after death, the longer we delay in making the analysis the greater is the amount of sugar there found. In fact, under certain circumstances, the duodenum of an animal, immediately after death, may contain only a trace of sugar, and yet an enormous amount be found in it in about half an hour later. The reason of this being, that so long as the animal lived the sugar was absorbed from the intestine by the veins and lacteals as rapidly as it was formed. Whereas, although when it died the absorption process became arrested, the transformation of the starch into sugar by the pancreatic juice still went on. I need not take up any more time by discussing this question here; but refer those specially interested in the subject to the published writings of Thudichum(a), Beale(b), and myself(c). I may only further remark that even supposing that *not a trace of sugar* could be detected in the liver immediately after death, it would be no more scientific evidence that the liver did not manufacture sugar during life, than the finding no urine in the kidneys after death would be that the normal function of these organs was not to eliminate urine. A few more substantial data than any hitherto brought forward are required, ere we can venture to accept of negative results in the place of positive facts.

The next point is, "What becomes of all the saccharine matter?" Bernard thought that it was burned off in the lungs, its chief office being to sustain the animal heat; but as Chauveau and I pointed out, such a doctrine is utterly untenable, seeing that when the experiments are properly performed there is almost as much sugar to be found in the blood returning from the lungs as in that going to them. Thus, for example, in a fasting animal I found that while the blood of right side of the heart contained 0.10 per cent. of saccharine matter, that of the left contained very little less—namely, 0.085 per cent.(d) In another experiment, performed upon a cat, the blood of the right side of the heart contained 0.18 per cent. of sugar, and that from the left an exactly similar amount; and in order to be certain that no mistake had been

(a) *Brit. Med. Jour.*, March 17, 1860.

(b) " Urine and Calculi." Second edition, p. 236.

(c) *Proc. Roy. Soc.* No. 38, p. 289.

(d) For the manner in which the experiments were performed, see the author's paper in the *British and Foreign Quarterly Review*, July, 1857, p. 201

made in the determination of the sugar by the volumetric process, I carefully collected the reduced oxide of copper, dried, and weighed it. The amount of precipitate from both bloods was identical.

When one takes into consideration the nature of the lungs, he cannot be surprised at these results; for the lungs are not laboratories, like the stomach, but merely an aggregation of little sacs whose function is purely physical—at least, in so far as respiration is concerned. The only "vital" offices they perform are simply those required for their own development and preservation. The absorption of oxygen and exhalation of carbonic acid would perhaps go on just in the same way if a piece of goldbeater's skin supplied the place of the air vesicles; for the blood arriving there for the purpose of arterialisation never leaves the capillaries to become in any way incorporated with their tissue.

Does the saccharine matter disappear, then, in the capillaries of the general circulation? It would appear so, for both Chauveau and myself have proved, by carefully executed experiments, that less sugar is to be found in the veins returning from a limb than exists in the arteries going to it. Thus, in the blood from the femoral artery of a middle-sized dog in full digestion I found 0.24 per cent., while that of the corresponding vein contained only 0.16 per cent. of saccharine matter. From its thus disappearing in the capillaries, one is naturally led to believe that it plays a part in the nutritive process. We know, indeed, that while bees have the power of transforming sugar into wax, man and other animals change it into adipose tissue. The negroes are said to become fat and lazy during the sugar harvest from sucking the fresh canes, and long before the days of Bantingism all Medical men were aware that babies fatten on sugar quicker than anything else. In 1856, while investigating the uses of sugar in the animal economy, I gave cats and dogs large quantities of sugar-candy along with their other food, and I invariably found that, although they all seemed to like the mixture and become fat and sleek upon it, a period at length arrived when they notably ceased not only to gain weight, but also to relish the food. In fact, at length, some of them would rather starve than touch meat mixed with sugar. It must be remembered, however, that I was giving them enormous doses—from two ounces to a quarter of a pound daily. When the sugar was given in moderation no such effects were observed.

The last point connected with the physiology of the glucogenic function which it is essential for us, as Practitioners, to understand, is the origin of the nerve-force which calls it into play. Bernard has shown that by dividing the pneumogastric nerves in the neck, the secretion of sugar is at once arrested, and that the application of galvanism to the upper ends of the divided nerves not only re-establishes the secretion, but, if the current be continued sufficiently long, augments it beyond the normal amount, so that animals thus operated upon not unfrequently become diabetic. On the other hand, the application of galvanism to the lower ends of the divided nerves is not found to be followed by any such result. These experiments clearly indicate that the nerve-force which excites the liver to secrete saccharine matter does not travel from the brain, through the pneumogastric nerves to the liver, but rather that the stimulus proceeds along these nerves to the brain, and is from thence retransmitted to the hepatic organ through some other nerve-chain.

The data yielded by other experiments, which it is at present unnecessary to recapitulate, induced Bernard to believe that in the healthy animal the reflex action which incites the glucogenic function originates in the stimulus given by the respired air to the pulmonary branches of the pneumogastric nerves, and that this stimulus is reflected from the brain along the spinal cord and splanchnic nerves to the liver.

The point of departure of the normal nerve force which calls into play the glucogenic function of the liver may, at the first glance, appear a matter of little moment; but when we consider that the secretions of organs increase in proportion to the amount of stimulus applied to their nerves, and that an excess of secretion, which not unfrequently constitutes disease, arises in many cases simply from an exaggeration of the normal stimulus, we shall at once acknowledge the importance of thoroughly understanding the physiological, before attempting to remedy the pathological condition of an organ. When an answer has been given to the query, " Where is the sugar secreted?" the question next in importance to the Physician is most assuredly, " By what means is the secretion excited?" A satisfactory answer to the latter question may not improbably furnish a guide to the successful treatment of a hitherto considered incomprehensible disease.

There is no difficulty in accounting for the liver being excited to secrete sugar when an irritation is applied to the pulmonary branches of the pneumogastric nerve; but we are not necessarily forced to believe that they are the branches which normally call into action this peculiar function of the organ referred to. Indeed, if such be in reality the case, how does it happen that while the respiration, and consequently the stimulus, continue at about the same rate during the entire day, the secretion, which is said to be the result of the stimulus, varies at different times? At one hour it is known to be exceedingly active; at another, somewhat later, almost dormant. Such a result has no parallel in any other organ of the body. A certain amount of stimulus, *cæteris paribus*, invariably calls forth a similar and definite amount of action; and upon what grounds are we warranted in considering the function of the liver an exception to the general rule? I need scarcely detain you at present with further arguments against what appears to me an untenable hypothesis, as I believe the results of the subjoined experiments clearly indicate, that *if* the pneumogastric is the nerve which carries the stimulus to the brain, to be from thence transmitted by the spinal cord and splanchnic nerves to the liver, the point of departure of the stimulus is most probably in the liver itself, and that the cause of the reflex action may originate in the stimulating effect of the portal blood upon the hepatic branches of the pneumogastric nerve. If, for example, the stimulating effect of the blood of the portal vein be imitated as much as possible by injecting into that vessel substances, such as alcohol, ether, chloroform, methylated spirit, or ammonia, the liver is excited to secrete an excess of sugar, and the animal operated upon is for a time rendered diabetic.(a)

The conclusion to which the results of the experiments led me was, that stimulants produce diabetes by exciting the hepatic branches of the pneumogastric to transmit an impression to the nervous centre, to be from these reflected to the liver, and thereby cause the increased secretion of saccharine matter; and if this explanation be the correct one, it is very probable that the normal secretion of sugar is caused by the stimulating effect of the nutritive materials in the portal blood. The following facts materially strengthen this view :—During the time of digestion the blood of the vena portæ must, of necessity, prove most stimulating, as it is then loaded with nutritive materials ; and this happens to be exactly the period at which the greatest quantity of sugar is formed. On the other hand, the blood of the portal vein of a fasting animal contains very little nutritive material; consequently during this period the quantity of sugar ought to be lessened, and this, in fact, is exactly what occurs ; for, in a fasting animal, the secretion of sugar has invariably been found to be at its minimum.

(a) For a full account of these researches, see an article by the author in the *Brit. and Foreign Quarterly Rev.*, July, 1857.

The important bearing of these physiological data will be more fully appreciated when we are considering the pathology of diabetes : for, as will be presently seen, they furnish us with a key to the rational treatment of some of the forms of saccharine urine.

Pathology.

We now come to what, at least as far as we, as practical Physicians are concerned, may be considered the most important part of our subject, namely, its pathology.   Diabetes happening, however, to be one of those intricate diseases leaving behind it no characteristic anatomical lesion by which a clue to its seat can be obtained, it is not in the dead house, but in the clinical ward and physiological laboratory that the nature and treatment of the affection are most successfully studied.   What has now to be said of its pathology and treatment will, therefore, be found greatly simplified by the knowledge we have already obtained of its physiology.   In fact, we shall soon find that the presence of sugar in the urine, like the discolouration of the skin in jaundice, is not of itself the disease, but merely the most prominent sign of several widely-differing abnormal conditions, the correct appreciation and understanding of which entirely depend on our acquaintance with its physiology.

Diabetes has, for example, been found to follow upon —
Injury to the head, with or without fracture of skull.
Clot in the pons varolii.
Softening of the base of the brain.
Abscess of the cerebellum extending into the fourth ventricle.
Tumour (the size of a nut) in the left lobe of the cerebellum.
Disease of the sympathetic nerve.
Tumour of the pneumogastric nerve.
Deposit of bony spiculæ in the falx.

Excessive brain work.

Intense grief.

Sudden mental shock.

Blow on the epigastrium.

Pregnancy.

Uterine disease.

Disordered digestion.

Exposure to cold, etc.

Even in some instances diabetes seems traceable to hereditary constitutional peculiarity, for not only is it occasionally met with in more than one member of the same family, but it has even been found present in parent and child to the third generation. Thus we find it related (a) that Dr. Mosler had a patient whose father, mother, and two sisters died of diabetes, and that within three weeks after the woman came to the Hospital, her son, aged 15, also presented himself, suffering from the same disease.

Seeing, then, that diabetes springs from a multitude of causes, and that the symptom of saccharine urine is not itself the disease, but merely the most prominent sign of the hidden complaint, we can easily understand how a variety of morbid actions, quite distinct from each other, and perhaps of a diametrically opposite character, may produce it. For example, if the normal stimulus of the liver be exaggerated, an abnormal amount of sugar will be secreted, and if the quantity formed be greater than the amount requisite to supply the wants of the system, the excess, which then acts towards the organism as a foreign body, will be eliminated along with the urine, and the disease diabetes mellitus established. On the other hand, the case may be reversed, and there may still exist an abnormal amount of sugar in the blood to be eliminated with the urine, notwithstanding that only the proper amount has been secreted by the liver. This, of course, must of necessity happen when

(a) *Brit. Med. Journ.*, Dec. 31, 1864.

from any cause the process of assimilation has been interfered with, and the body fails to consume its normal quantum of saccharine material.

Hence, it is evident that we may have two perfectly distinct forms of the same disease, one of which might be named *Diabetes from excessive formation;* the other, *Diabetes from defective assimilation* (mal-nutrition).

I shall presently have occasion to show that this theory is equally supported by clinical observation as by the results of scientific inquiry. It will be found, too, that it furnishes us with a logical explanation to the fact which is so familiar to all who have had much experience with diabetic cases—namely, that while one man improves and gains both weight and strength on animal diet, another on the same regimen loses both flesh and energy, notwithstanding a diminution, perhaps, taking place in the amount of sugar excreted.

Piorry's idea of curing diabetes by giving, instead of withholding, saccharine matter may after all turn out to have more reason in it than at first sight appears. Like the animal diet system, however, it will be found only applicable to a certain class of cases. The great point for us, then, is to be able to discriminate between the two forms of morbid action, in order that our plan of treatment may be the one best adapted for the particular case.

How can this be done? Fortunately, the two forms of diabetes have certain peculiarities connected with their history which enables us in most cases to distinguish the one from the other.

In those resulting from *excessive formation* the patient is not necessarily emaciated and weak. He may, on the contrary, look both fat and ruddy—appearing, in fact, to be in the very bloom of health. This is especially true in slight cases, as well as in the early stage of even those which run on to a fatal termination. In this class of patients it is not until the disease has made considerable inroads on the constitution that there is any marked emaciation.

In the second class of cases, on the other hand, namely, those resulting from *defective assimilation*, emaciation is one of the earliest and most prominent symptoms, the loss of flesh being often very marked before the nature of the disease is detected. Exactly the reverse of what frequently occurs in the first class, in which the patient occasionally receives the first warning of the affection from the accidental detection of sugar in his urine.

An inordinate thirst and excessive elimination of urine is in all cases an indication that the disease is already in its second stage, the first stage being indicated, in those arising from *excessive formation*, by saccharine urine alone, and in those from *defective assimilation* by saccharine urine coupled with loss of flesh.

At one time, when a patient was said to have diabetes, it was considered tantamount to saying his days were numbered. As our knowledge has increased, however, we have learned that although it is but seldom possible for us to eradicate the disease, we can, nevertheless, so mitigate its effects as not only to prolong the life of the individual, but to render it a tolerable, if not even an agreeable one. Prout used to declare that diabetic patients are for ever standing on the brink of a precipice, for they are in constant danger of succumbing from the indirect as well as the direct effects of the disease, in consequence of there being in them an absence of that vital stamina which enables healthy persons to resist the inroads of disease. This is, no doubt, true, for the mal-nutrition which diabetes induces renders the constitution prone to take on low inflammatory actions, which, if set up in a healthy individual, would produce but a temporary illness, but when engendered in the diabetic, rapidly run on to a fatal termination. But, as forewarned is forearmed, this need not distress, but only make

R

us the more careful at once to remedy any trifling deviations from the normal standard, which, in the healthy, might be allowed to pass unheeded.

Having only specified two forms of diabetes—1st, That by *excessive formation ;* 2ndly, That by *diminished assimilation* — it may naturally be asked how this statement can be reconciled with the one previously made regarding the multitude of causes inducing the disease? Easily enough, as we shall now see.

Nerve Lesion and Disease as a Cause of Diabetes.

It is generally admitted that the various secreting organs of the animal body are stimulated to perform their different functions by direct and reflex nerve-action, and the liver proves to be no exception to this law. If the pneumogastric nerves be divided in the neck, the glucogenic function of the hepatic organ is arrested, and when galvanism is applied to the upper (but not to the lower) ends of the divided nerves the saccharine secretion is re-established. Moreover, it has been ascertained that this reflex nerve-force, which travels by the pneumogastric nerves to the brain, is from thence reflected along the spinal cord, splanchnic nerves, and solar plexus, to the liver. (Bernard.) It has been further ascertained that it is not at all essential that the galvanism be applied to the cervical portion of the pneumogastric in order to re-excite the saccharine secretion, for, according to another well-known physiological law, irritation produced at any point of a reflex nerve circuit is invariably followed by the same phenomena.

As what I am now attempting to explain is the keystone to the comprehension of the pathology of diabetes by nerve influence, it is absolutely essential that my meaning should be clearly understood. I must, therefore, solicit your attention for a moment to the following experiment :—

Let the hind legs of a frog be prepared in the manner represented in the accompanying woodcut, every attachment being severed except that of the sciatic nerves with the spinal cord. Besides this, let the motor roots be divided on the right side (Fig. 27, *a-b*), and the sensory roots on the left (*c-d*), so as to

FIG. 27.

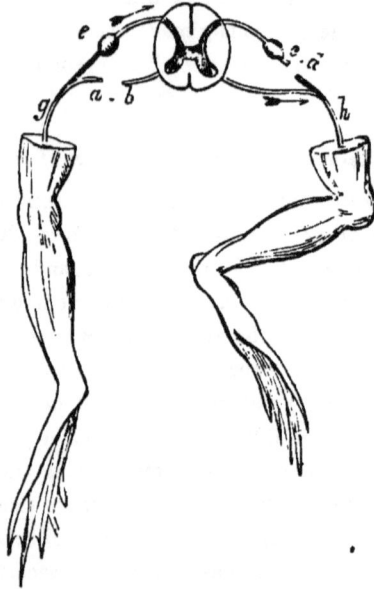

G. 27.—Reflex circuit in hind legs of frog attached to the spinal cord. *a-b*, anterior motor root of (*g*) right sciatic nerve. *c-d*, posterior sensory root of (*h*) left sciatic nerve.

leave only one complete nerve-chain. The application of a slight galvanic current to the toes of the right limb will then be instantly followed by contractions in the muscles of the left, showing that the nerve stimulus is transported through the sensory nerve of the right limb to the spinal cord, and reflected from thence, through the motor nerve, to the muscles of the left extremity. In order to produce this muscular movement, however, it is not at all necessary tha the galvanism be applied at the extremity of the sensory nerve; for irritation produced in any part of its course (*g*, for example), at its root (*e*), or even in the opposite nerve chain (*h*), will still be followed by the same muscular contractions. Now, it so happens, with regard to the glucogenic function of the liver, that the pneumogastric is the analogue of the sensory, and the spinal cord and splanchnic of the motory nerves of the frog's limb; so that an irritation applied to any part of the pneumogastric nerves, spinal cord, or sympathetic reflex glucogenic circuit, is as effectually followed by a secretion of sugar as the application of galvanism to the sciatic nerves of the frog's limb is followed by muscular contraction. Moreover, the irritation, galvanic or other, has only to be applied to the glucogenic circuit sufficiently long or sufficiently powerfully in order to induce saccharine urine. Hence it is within the power of the physiologist to render animals artificially diabetic by a variety of experimental procedures implicating the various parts included in the reflex nerve-chain above indicated. Thus diabetes can be artificially established by applying the irritation to the extremities of the pneumogastric in the liver; as when stimulants are injected into the portal vein (the author), or by irritating the pulmonary branches distributed in the lungs, as where chloroform is inhaled (Reynoso), or by applying galvanism to the cervical portion of the nerves, or by irritating their roots, as when injury is applied to the floor of the fourth ventricle (Bernard), or by acting on the downward chain, as, for example, when the splanchnics are divided. (Graffe.) In fact, I might go on to prove experimentally that the *rationale* of saccharine urine, by reflex nerve action, is as comprehensible as the *rationale* of the muscular contraction in the frog's limb. Nay more, just as we may have muscular contraction following upon the *indirect* application of a stimulus to the nerves in the frog's limb, so we may have saccharine urine following upon the indirect application of the stimulus to the glucogenic circuit. For example, diabetes may be induced by a blow on the epigastrium, or from an injury to the head, as in the case which I already cited in a previous lecture. The patient at the time he was admitted into University College Hospital under the care of my colleague, Professor Erichsen, was suffering from paralysis consequent upon injury to the head received in falling from a scaffold. It may be mentioned that it was in consequence of the man having fallen upon the occipital region that I suggested the possibility of his being diabetic, and had the urine tested. In this case sugar was found the very first time the urine was examined, so that it is impossible to say when it first became saccharine; no sooner was the sugar detected, however, than the nurse was ordered to collect daily all the urine voided during the twenty-four hours, and transmit it to the laboratory for quantitative analysis.

The analyses were made by my former pupil, Dr. Pringle, and the following is a table of the results :—

1857.	Quantity of urine.		Quantity of sugar.	
	In cub. centim.	Oz.	Grammes.	Grains.
July 18th	952 =	30·7	2·38 =	36·89
20th	1000 =	32·2	5·00 =	77·50
21st	2187 =	70·5	6·38 =	98·89
22nd	1875 =	60·5	3·90 =	60·45
25th	1955 =	63·6	3·48 =	53·94
28th	1000 =	32·2	1·25 =	19·37
31st	Not sufficient sugar for quantitative analysis.			
August 3	No sugar.			

It will be seen that at first after the receipt of the injury, the sugar went on steadily increasing, and then, as the patient recovered, it as gradually steadily diminished, until at length it entirely disappeared.

This was prognosed when I first saw the case   For the

condition of the man, as far as his diabetes was concerned, was exactly similar to the state of rabbits rendered artificially diabetic by puncture of the fourth ventricle, and we all know that at first the amount of sugar gradually augments until the injury in the cerebral substance begins to heal, and then the diabetes as gradually disappears.

Some may say, we had no proof beyond the existence of saccharine urine that injury had occurred to, or in the neighbourhood of the fourth ventricle. This is very true, but I think the reasoning from analogy justifies the diagnosis. However, as some may be sceptical on this point, I shall now cite a case where the diagnosis may be said to have been verified by the post-mortem examination.

On November 13, 1859, a man was brought to University College Hospital in a state of insensibility, with slow, stertorous respiration, a weak pulse, and contracted pupils. As the patient died without ever regaining consciousness, the only history of the case that could be obtained was that he was seen suddenly to fall down in the street, and when picked up was in a state of complete insensibility. At the autopsy, the only lesion found was a clot encircling the medulla oblongata and extending upwards into the pons Varolii, its situation being apparently such as to have caused considerable pressure on the fourth ventricle and roots of the pneumogastric nerves. The presence of sugar in the urine was sought for, and in this case was demonstrated without the least difficulty. Although no quantitative analysis was made, there can be no doubt that the sugar was abundant, as the urine became quite brown on being boiled with potash, and freely reduced the oxide of copper. The flow of urine must have been also copious, as the bladder was found greatly distended by it after death. The substance of the liver contained a large amount of sugar. and even the brain itself was impregnated with the saccharine matter, for an aqueous decoction yielded all the reactions in a very decided manner. Now, unless we are to suppose that this patient had been previously suffering from diabetes, of which his body presented none of the external signs, one is led to the conclusion that the saccharine urine was the direct result of the pressure of the clot on the fourth ventricle.

Another case of diabetes from external injury to the head may be mentioned, as it, like the one already cited, affords an example of spontaneous cure. The urine of a woman, under the care of Dr. Todd, who had hemiplegia and rigidity of the right side, in consequence of falling down stairs upon her head, when examined on the 20th day after the injury was found to be of a specific gravity of 1021, and to contain a small quantity of sugar, which gradually disappeared as the patient got well.

Several years ago, Dr. Goolden had a case of a similar kind in a railway stoker who was struck on the occiput by the handle of a crane, and remained diabetic during his illness.(a) Some nerve lesions give rise to a diabetes of a still more permanent character, as the following case will show. Dr Ramsbotham, of Amwell-street, informs me that when on a visit to a Medical friend in Kent a year or two ago, he found him in a fair state of health, with the exception of feeling extreme thirst, and suffering from great languor. Suspecting diabetes, he examined the urine, and found it to be above the average in quantity, of a specific gravity of 1044, and loaded with sugar. The gentleman had never suspected that he was diabetic; it is therefore impossible to say how long this state of matters had existed. The important part in the history of the case is, that four or five years previously he had been thrown from a horse, and in some way or other got his head forcibly bent forwards, and for a length of time afterwards he could not turn it to either side without great pain; at the same time both the senses of smelling and hearing were

(a) *Medical Times and Gazette*, May 15, 1858.

blunted. The last report Dr. Ramsbotham had of the case was that " the diabetic symptoms were ameliorated, or at any rate, had ceased to be a source of annoyance."

While calling attention to these cases of diabetes, the result of cerebral lesion and disease, I must not omit to mention that an inordinate flow of urine may follow upon concussion of the brain without being associated with the presence of sugar. An interesting case of this kind was observed in M. Moutard, Martin's wards, in which, after a concussion of the brain, followed by hemiplegia, the patient suffered from extraordinary polydipsia. Under the influence of a seton, the cerebral symptoms amended and the polydipsia entirely disappeared.(a)

As cases of polydipsia are not common, and some authors have even entirely doubted their existence, I may mention an interesting case related by Dr. Watson :—A boy, aged 11 years, who was very thin, but not unhealthy looking, was troubled with thirst and frequent micturition, passing daily nine or ten pints (180 to 200 ounces) of pale dilute urine, of a specific gravity of 1002—sometimes, indeed, scarcely heavier than distilled water. Every remedy was tried, but he left the Hospital no better than when he entered it. Even when his drink was limited to a pint and a-half, he still passed ten and a-half pints of urine.(b)

In a case of my own, a man, aged 57, passed, on an average, eight pints (4960 c.c.) of urine of a specific gravity of 1010 in twenty-four hours, the daily average of urea being 70 grammes (1085 grains). Notwithstanding that the patient passed so much urine, he only drank about half that quantity of liquid, and, strange to say, although he had an exceedingly dry skin, he never complained of thirst. The polydipsia had lasted three years, and during that time he lost forty-two pounds in weight. No sugar was ever detected in his urine.

(a) *Gazette des Hôp.*, No. 18.

(b) Lect. on the Practice of Physic, vol. ii., p. 611.

Cerebral disease—not consequent on injury—may also be productive of saccharine urine. Thus, for example, epilepsy is occasionally accompanied by diabetes.

In a case of this kind which occurred in the Hospital for the Epileptic and Paralytic, under the care of Dr. Brown-Séquard, the pons Varolii and calamus scriptorius of the fourth ventricle were found diseased.(a) The man, aged 53, whose urine was of a specific gravity of 1037, and "contained a. good deal of sugar," had increased sexual desire before death.

A short time since I examined the urine in a case of Dr. Hughlings Jackson's, in which, with disease in the near neighbourhood of the apex of the calamus scriptorius, there was for several months sugar in the urine.

M. Luys related to the Paris Anatomical Society a case of diabetes which lasted over three or four years. At the autopsy the anterior wall of the fourth ventricle was found highly vascular and its consistence notably diminished, very slight scraping bringing off a gelatinous pulp of a brownish yellow colour, the ventricle exhibited the colour very remarkably at certain spots. The lesion consisted in the molecular destruction of the histological elements, and their *débris*, loaded with yellowish granulations, gave rise to this peculiar colour. This lesion may be regarded as exactly corresponding to the traumatic one produced by the experimental physiologist which leads to an exaggeration of the glucogenic function of the liver and the consequent passage of sugar into the urine.(b)

In other cases of cerebral disease diabetes also occurs. Thus, Laudet (c) mentions the following four cases in which he observed it. The first is that of a woman, aged 32, who while pregnant was affected with the loss of vision of the left

(a) Lockhart Clarke in Beale's *Archiv.*, vol. iv., p. 146.
(b) *Bulletin* de la Soc. Anatom., voL v., p. 219.
(c) Acad. des Sciences, March 2, 1857.

eye, without paralysis. She at the same time suffered from vomiting and headaches, and on one occasion had symptoms of coma accompanied by paralysis of the third and fifth part of cerebral nerves. These symptoms were accompanied with great thirst and the other signs of diabetes mellitus. Under the administration of iodide of potassium the paralysis and diabetes diminished.

The second was that of a woman, aged 53, suffering from hemiplegia and epilepsy, supposed to have originated in apoplexy. Two years later diabetes appeared, and in one year more also albuminuria.

The third was that of a woman, aged 80, suffering from hemiplegia of the left side. In eight months' time she had great thirst consequent upon an attack of diabetes. This was followed by gangrene of the right foot, from which she died.

The fourth was that of another woman, aged 39, who in the sixth month of gestation became paralysed, with convulsions. Gradually these disappeared, and only slight swimming in the head remained. Six years afterwards she was found to be diabetic. This patient ultimately died of small-pox.(a)

Over-mental exertion I have put among the causes of saccharine urine. An interesting example of this kind has recently fallen under my notice in the case of one of our students. During three weeks of mental excitement, sugar was daily detected in his urine. After a fortnight's rest, it entirely disappeared, but recurred a few days after the resumption of the mental labour, again to disappear almost as soon as the labour was discontinued. I have seen again and again mental labour increase the sugar in diabetic patients. In the case of a clergyman, to which I shall subsequently have occasion to refer, on several occasions, I found the daily amount of sugar double itself when he was preparing for his College

(a) *Gazette Médicale*, March 14, 1857. For some other cases resulting from cerebral lesions and disease, see *Gazette Médicale de Paris*, March 20, 1858.

examinations, and again gradually return to the usual average almost as soon as they were passed, without either medicine or diet being altered.   •

Irritation of the pneumogastric or of the sympathetic nerves may give rise to diabetes.

Mr. Nyman relates the case of a Professional gentleman who had long suffered from diabetes mellitus. "After we had examined," he says, "the above-named organs (chest, abdomen, and brain), we commenced the dissection of the nervi vagi. On the left nothing abnormal could be found, but on examining the right, we met in the thorax immediately behind the bifurcation of the bronchii a calculous mass (tumour) the size of a hazel-nut, under the entire trunk of the nerve, and appeared to have exercised considerable pressure upon it." He adds that it is the third time that he has met with this morbid condition in persons who have died from diabetes. The urine of this patient contained 6 or 7 per cent. of sugar. (a)

One case is also recorded in which disease of the sympathetic caused diabetes. In a report of the cases occurring in the Royal Infirmary, Edinburgh, under the care of Dr. Duncan, it is noticed that in one case of diabetes there was found, after death, to be an hypertrophy of the lumbar sympathetic. (b)

Saccharine urine may also arise as the result of indirect nerve irritation; thus it is met with in tumours of the abdominal organs and in pregnancy, in which cases the only way to account for it appears to be by the irritation caused by pressure or otherwise of the abdominal sympathetic, perhaps the cœliac plexus or some of its branches.

For further information regarding the effects of nerve lesions and irritation in the production of diabetes, reference must be made to the physiological views previously expressed.

(a) *Dublin Hosp. Gaz.*, from *Swedish Trans.*, July 15, 1857.
(b) Longet, vol. ii., p. 635.

We now come to the consideration of

Disorders of the Digestive Function as a Cause of Diabetes.

And, first of all, we must notice those cases which arise apparently from the use of stimulants. It is a well-known fact that cases of diabetes are much more common in Great Britain than on the Continent, and I, like many others, attribute this to the fact of a much greater quantity of alcoholic drinks being used among us than among other European nations. A few years ago (August, 1861), while on a visit to Munich, in the course of conversation, Dr. Pfeüfer, the Professor of Clinical Medicine, told me that he had only had one case of diabetes in the Hospital during the whole six years he had been in Munich; whereas, while Professor in Heidelberg, although the Hospital was much smaller, he had on an average four or five diabetic cases in the course of each year. Now, the only way to account for this difference was that the Munich Hospital draws its supply from a beer (Bavarian) drinking district, while the Heidelberg Hospital obtains its supply from the wine-drinking districts of the Rhine, etc. And, curious enough, it turned out that even the single case of diabetes Pfeüfer had in Munich, he added, came from the Palatinate.

There can be no difficulty in explaining this action of stimulants, for, as said in the physiological part of the lecture, diabetes can be artificially induced in healthy animals by injecting small quantities of alcohol, ether, chloroform, or even ammonia, into the portal circulation.

For example, I injected 10 cubic centimetres of sulphuric ether diluted with 30 cubic centimetres of water into one of the branches of the portal vein(a) of a full-grown Newfound-

(a) The experiment is very easily performed by using a sharp-pointed syringe, which can be pushed with facility through the coats of one of the large mesenteric veins. Care must be taken not to employ too much of the stimulant, as it is only up to a certain point that it possesses the power of producing artificial diabetes. If too large a quantity be injected, it seems rather to destroy than increase the glucogenic function.

land dog half an hour after he had been fed. When he rose
up after the operation, he appeared intoxicated, and staggered
a little as he moved about. This effect, however, soon disappeared, and in a few hours the animal looked as if nothing
had been done to him. In two hours after the injection was
made, I passed a catheter into his bladder; but did not obtain
sufficient urine to enable me to satisfy myself whether it contained sugar. Some hours afterwards, when I had obtained
enough urine, I found that it readily reduced the copper in
Barreswil's liquid, thus indicating the presence of saccharine
matter. To assure myself that this effect was not due to the
presence of any other substance, I boiled the urine in order
to coagulate the albumen, of which it contained a little, then
evaporated it almost to dryness, dissolved the residue in boiling alcohol, and filtered. The filtered liquid was next evaporated to drive off the alcohol, and an aqueous solution made.
On testing the latter for sugar with the sulphate of copper
solution, its presence was clearly indicated. Although by
this method the existence of saccharine matter was rendered
almost undeniable, I still wished to convince myself of its
presence by some other means. The urine which the dog
passed the next day was therefore fermented, and carbonic
acid gas and a trace of alcohol was obtained, thus placing
beyond a doubt the existence of sugar.

The following case shows that ammonia has the same power
as ether in causing the liver to secrete an abnormal amount of
saccharine matter :—

Into the portal vein of a good sized dog in full digestion, I
injected fifteen drops of liquor ammoniæ diluted with forty
cubic centimetres of water. In twenty hours afterwards, on
the animal being killed, his bladder was found enormously
distended with urine, which not only reduced the copper in
the liquid of Barreswil, but fermented rapidly.

These two experiments are selected from a number of others,

some of which I had the honour of performing in 1853 at the
College of France before a Commission appointed by the
Société de Biologie; but as the results then obtained are
identical it is unnecessary to cite them. I cannot refrain from
mentioning with what pleasure I perused a communication of
Bernard's, entitled " On the Influence of Alcohol and Ether
on the Secretions of the Digestive Canal, of the Pancreas, and
of the Liver," read before the Société de Biologie.(m) M.
Bernard, instead of putting the alcohol and ether, as I had
done, directly into the portal vein, introduced them, by means
of a long œsophagus tube, into the duodenum of dogs, and
allowed them to be absorbed through the walls of the intestine, into the portal circulation. The result, as might *à priori*
have been anticipated, was identical with what I had previously obtained. Bernard, in fact, found that six centimetres
of alcohol mixed with an equal amount of water, was sufficient
to excite the liver to secrete a large quantity of sugar, even
while the animal was fasting. With ether employed in a
similar manner, he obtained no less successful results.

Rosenstein(n) has still further confirmed the value of the
physiological data I obtained, by a very complete series of experiments on the influence of different kinds of drink upon the
quantity of sugar, salt, and urea, daily eliminated by a diabetic
patient. He performed the experiments, he says, with the
view of ascertaining if the conclusions I arrived at were
equally applicable to the human subject as to the animals
operated upon, and he came to the conclusion that they are
identical from his having obtained the following results :—1.
Coffee, while it diminished the elimination of urea, increased
that of the chloride of sodium and sugar. 2. Bavarian beer
had a precisely similar effect. 3. The influence of wine was
also similar—namely, to decrease the elimination of urea, and

(m) *Gazette Médicale de Paris*, May 10, 1856.
(n) Virchow's *Archiv.*, 1858, p. 461.

augment that of the chloride of sodium and sugar. Lastly, he found that vinegar exerted a similar influence on the elimination of these substances.

If any one doubt the truth of the assertion that stimulants excite diabetes, let him select a case of that form of the disease arising from excessive formation, and after having carefully estimated the daily amount of sugar eliminated by the patient, allow him to drink a few glasses of wine during the next few days, and watch the result. He will soon find that the ingestion of stimulants is followed by an increase of sugar, and if stimulants increase the amount of saccharine matter in the urine of the diabetic, we can easily understand how their excessive use may induce the disease in individuals *predisposed* to it. The effect of stimulants introduced into the portal circulation also explains to us how a disordered digestion is not unfrequently followed by saccharine urine. I may here relate a curious fact in illustration of the latter remark. In 1852, at a time when I was much occupied in studying the physiology of diabetes, I regularly tested my urine twice a-day, and on one occasion I found it to contain a small quantity of sugar. On the day in question I had partaken freely of asparagus salad; and, thinking that this might perhaps be the cause of the presence of the sugar, I determined to try the effects of a greater quantity. The following day, the sugar having entirely disappeared from the urine, I again partook of the same kind of salad, both in the morning and afternoon. In the evening, on testing the urine, I found very distinct indications of sugar. As the observation was to me one of great interest, I determined to make some further experiments on the subject, in order to discover how many hours this state of saccharine urine would continue. During two days I ate large quantities of asparagus salad, taking care to have it made as stimulating as possible with vinegar and pepper. The result was far beyond my expectations, for instead of the sugar disappearing from the urine in a few hours after I had ceased partaking of the diet in question, it continued to be secreted during several days, until I at last became very much alarmed lest the disease had been permanently induced. On the evening of the fourth day the sugar had almost entirely disappeared, but on the fifth it returned in increased quantity, so much so that a drop of urine falling on the boot left a distinct white spot. I could not account for the recurrence of the disease, as I had been particularly careful in my diet during the two previous days.

I have mentioned this experiment because it appears to me that if a flow of saccharine urine be induced in a healthy person, as I consider myself to be, by disordering the digestion and over-exciting the liver, it is very probable that a cause insignificant in itself, but operating upon a predisposed con· stitution, may tend to produce the disease. Sugar in the urine has been found after eating cheese and other indigestible substances. It is worthy of remark that Dr. Jessen, of Dorpat, has rendered horses diabetic by feeding them with hay damaged by moisture. M. Léconte has also found sugar in the urine of dogs after he had administered to them the nitrate of uranium. Several other substances have the same effect, and I have no doubt but that a greater number of stimulants will be afterwards found to produce similar results.

We now come to the consideration of the second form of diabetes — namely, that in which there is no proof of the existence of an *abnormal production* of sugar by the liver, but in which there rather appears to be a *diminished consumption* of the amount normally produced. In illustrating this point I have to rely entirely on clinical data for the opinions at which I have arrived.

It will be remembered that several years ago Piorry communicated to the French Royal Academy a paper, in which he stated that he had successfully treated cases of diabetes by giving the patients sugar-candy, and that he was led to do so from his believing that the cause of death in diabetes is the loss of the sugar which is so necessary to support life. The cane-sugar was therefore given in order to supply the place of the sugar that was lost.

Not long after reading these views, when on a visit at Bristol, Dr. Budd showed me a young man whom he was treating by means of saccharine diet, with apparent benefit. Subsequently, Dr. Sloane(a) and Mr. Amyot(b) published papers on the saccharine treatment of diabetes, expressing similar opinions. On reflecting over the views promulgated by these gentlemen, and comparing them with the results of my own clinical and physiological experience, it gradually occurred to me that there might be two distinct forms of the disease, such as I have already described under the head of diabetes by *excessive formation*, and diabetes by *diminished assimilation;* and further investigation has not only confirmed this view, but enabled me to discriminate between the two classes of cases by the history and appearance of the patients alone. Before giving examples of each form of the disease, let me first explain one or two things regarding the symptoms of diabetes, which, I fear, are but imperfectly understood.

First, as regards the amount of urine eliminated, some appear to consider it a most important sign, and one which we should try and check as soon as possible. Now, I beg to differ from them, for I believe it fortunate that the diabetic patient does pass an excess of water.

The excessive elimination of sugar is not consequent upon the increased flow of urine, for we may have, as already seen, an excessive flow of urine without sugar,—but exactly the reverse; the excessive flow of urine is consequent on the elimination of the sugar. The sugar, in order to be eliminated, ·

(a) *Brit. Med. Jour.*, March, 1858.
(b) *Med. Times and Gaz.*, March, 1861.

s

must be dissolved, and, in order to be dissolved, must have water, and the more water the more readily does the elimination of the sugar take place. Some may say, we want to stop the elimination of the sugar. Not so; we want to stop the disease inducing it,—not the elimination of the sugar, which is the mere result of the disease. Retaining the sugar in the blood would only tend to hasten the death of the patient by still further deranging the nutritive functions, and causing an abnormal diasmose by altering the relative specific gravity of the blood and other secretions. Remove the cause of the accumulation of saccharine matter in the blood if you can, but if you cannot do that, aid, instead of trying to retard, its elimination from the body.

Diabetic patients generally pass more liquid than they take—about one fifth or one-quarter more—and although they ought never to drink more than they feel a want for, yet they must never be stinted, for their continual thirst is but nature's cry for relief. If the patient did not drink, the blood would soon get too thick to circulate freely through the vessels, and a variety of secondary diseases would be induced. Stopping the drink diminishes the elimination, but does not stop the formation of the sugar. When the formation of sugar decreases, the urine of its own accord becomes diminished.

As regards the amount of sugar eliminated, which is generally considered an infallible criterion of the condition of the patient, experience has shown me that such an idea is erroneous. A diminution in the sugar in cases from *excessive formation* is always a good sign, but in those arising from diminished assimilation it is, on the contrary, occasionally the reverse. Thus I have again and again seen patients who were gradually succumbing from the starvation effects of a restricted diet, at once pull up and improve on a judicious mixture of vegetable food, notwithstanding that the amount of sugar in the urine was thereby greatly increased. It is the weight of the patient, not the quantity of eliminated sugar, on which we ought to rely in such cases.

Although advocating the employment of vegetable diet in cases of diabetes of the second class, I do not wish it for a moment to be supposed that I agree with Piorry in thinking that the cause of death, even in this form of diabetes, arises from the loss of sugar ; for, on the contrary, I think it springs from the inability of the body to assimilate the sugar it possesses. In such cases, therefore, I give vegetable food, not because it contains sugar, but because it possesses many of the other substances necessary for the purposes of nutrition, which neither exist in the same quantity nor in so easily an assimilated form in animal diet.

It is well known from the reports of travellers among savage nations, that men restricted solely to animal diet must consume an almost fabulous amount in order to obtain sufficient of all the ingredients requisite in the processes of life. We know, too, that an animal can be most effectually starved by limiting him to one particular element of food, although that element be even albumen.

The benefits derivable from Piorry's plan of treatment, therefore, in my opinion, arise from the fact that when he gives sugar he at the same time ceases to restrict the patient to animal diet, and that in the mixed food they find many of the materials essential to life much more abundantly and in a more easily assimilated form than in animal diet.

I shall now give a few typical cases illustrative of the two principal forms of diabetes.

Diabetes from Excessive Formation.

In the beginning of 1860 a young gentleman, aged 19, suffering from diabetes, was brought to me by his brother, a Medical Practitioner, who had detected the disease two years previously. This patient had already been under various systems of treatment. What appeared to agree with him best was animal diet, coupled with small doses of chlorodyne. To look at the patient one would have thought that he was a perfectly healthy individual. His weight was 135 lbs. ; the appetite was moderate ; and the amount of urine passed was not at that time excessive. The object of bringing the patient to me, it appeared, was in order that I might, if possible, suggest some remedy to replace the chlorodyne, the consti-. pating effects of which were anything but agreeable. On carefully inquiring into the history of the patient, the case appeared to be one of diabetes by excess, and the origin of the mischief could in some measure be traced to some irrita- tion in the liver, which was painful at its lower margin, the pain being much increased on pressure. Having an intelligent Practitioner to deal with, I at once gave my view of the case, and explained how, as scarcely any two cases of diabetes are precisely alike, it would be necessary to try the effects of different forms of treatment in order to discover what would be best for this particular case.

The following table is an abstract of the results :—

Date.	Quantity.	Sp. gr.	Sugar.	Urea.	Weight.	
1860. March 27.	1980	1026	grms. grs. 99·05=1535·27	grms. grs. 45·50=705·25	lbs. 135	Animal diet; ordi- nary bread ; 5—6 glasses of wine and chlorodyne.
April 10 .	1680	1034	129·24=2003·22	36·96=572·88	—	On the same diet and drink as above, but no chlorodyne.
April 20	1680	1034	93·33=1446·61	50·40=781·20	—	Animal diet ; gluten bread and wine.
May 13	2085	1024	79·49=1232·09	57·26 887·53	—	Animal diet and gluten bread, without wine or medicine.

Date.	Quantity.	Sp. gr.	Sugar.	Urea.	Weight.	
August .	2010	1023	87·39=1354·54	46·33=718·11	—	Animal diet and gluten bread, one glass of wine, and ext. conia.(a)
December 1861.	1860	1022	90·00=1395·00	—	149	Same diet and treatment as last.
January .	1798	1024	119·20=1847·60	35·96=557·38	154	Diet the same as before, but part of the conia (6 grs.) replaced by ¼ gr. of belladonna.
February.	1736	1026	157·82=2446·21	34·70=537·85	—	On ordinary diet and without medicine.
March .	1798	1037	224·66=3482·23	46·51=720·90	—	Still on ordinary diet and without medicine.
July .	1675	1031	83·75=1298·12	50·25=778·87	—	On animal diet, gluten bread, and cannabis indica.
November	1829	1026	61·30= 950·15	25·74=398·97	—	Diet as before; medicine a mixture containing conia, cannabis indica, and hydrocyanic acid.
December 1862.	1736	1028	49·91= 773·60	39·92=618·76	—	Do., do.
July .	1736	1030	57·60= 892·80	58·31=903·80	—	Do., do., but less restricted diet.

The average amount of sugar passed by this patient during the next six months being from fifty to sixty grammes (775 to 930 grains) and his weight 156 pounds, a quantitative analysis of the sugar was no longer thought necessary. I may add that when I last saw the gentleman in August, 1864 (on the occasion of his bringing to me a poor lad, whose case I shall presently relate), he looked in excellent health, being, as he said, without any feelings of discomfort, although he had still to continue his medicine, for as soon as he neglected it, the sugar again increased.

(a) As many of the preparations of conia are utterly valueless, in consequence of their containing little or none of the active principle, that given was first tested on frogs before being employed.

The next case is that of a lady who also suffers from diabetes by excess. Before coming to me she had been under a gentleman connected with one of the City Hospitals, who treated her according to what has been hitherto considered the orthodox principles of restricted diet, upon which treatment she had thriven so well that when she walked into my consulting-room I never dreamt that she came to talk about herself, but imagined she came about the health of another patient with whom she had been to see me a short time previously. She looked, in fact, the very picture of health ; being plump, hale, and rosy.

Like many patients in her position of life, she thoroughly understood the nature of her disease, and criticised the opinions of our first authorities on the subject in a manner which somewhat surprised me—especially when she finished by saying she had come to put herself under my care, not for the purpose of being dieted, *but for the purpose of being able to live without being dieted.* She was tired, she said, of gluten bread, etc., and wanted to live like other people. She had brains enough to see that dieting kept down the sugar, not by curing the disease, but merely by stopping the supply, and that as soon as she gave up the restricted diet, back came all the disagreeable symptoms.

On analysing the urine I found she was in as good a condition as restricted diet could make her ; the quantity of urine passed was moderate, of a specific gravity of 1030, and the amount of sugar comparatively trifling,—310 grains. Diagnosing the case to be one of diabetes by excess, I honestly told her that although her case was one of the most favourable as far as longevity was concerned, it was one of the least satisfactory kind to put on ordinary diet. At the same time adding, that the treatment must, in the first place, be entirely experimental, it being impossible to say what remedy would most successfully allow her to dispense with dieting.

The case so closely resembled the preceding in its general characters, that I ventured on the same line of treatment, and, as will be seen by the subjoined table, with a favôurable result. The sugar gradually diminished, until at length, after seven months' treatment, the saccharine matter had entirely disappeared from the urine.

Being too sanguine as to the result, in an unlucky moment I consented to the patient's throwing aside the restricted diet, and whether this was done too suddenly or not I cannot say, but certain it is, that before a month was over, the urine contained as much, and even more sugar than when first I took the case in hand.

	Quantity.	Sp. gr.	Sugar.	Urea.	
1863.			grm. grs.	grms. grs.	
December	1860	1030	20·00=310·0	—	On animal diet and gluten bread; no other treatment.
1864.					
January .	2480	1022	5·00= 77·5	81·84=1268·52	After taking conia and cannabis Indica during fourteen days; animal diet and gluten bread, as before.
March .	1830	1030	4·23= 65·5	78·56=1217·68	Do., do.
April .	1674	1023	2·10= 32·5	50·22= 778·41	On the same diet, but, after increasing the narcotic by one-third (40 ℩), during last three weeks.
May .	1395	1030	Too little to calculate	69·75=1081·12	Same as last, but has been occasionally indulging in a very small piece of ordinary bread.
June .	1178	1030	Only a trace	44·76= 693·78	Diet still restricted ; has once or twice indulged in a potato at dinner; takes the medicine as before.
July .	1364	1016	*None*	39·55= 613·02	Has had one meal a-day of ordinary diet during last ten days.
August .	1984	—	20·60=319·3	59·50= 922·25	After being for nearly a month on ordinary diet.
November	1550	1030	13·00=201·5	74·40=1153·20	Restricted diet and medicine as at first.
December	1488	1026	10·00=155·0	52·08= 807·24	Do., do.

One very remarkable peculiarity of this case is the very large amount of urea passed by the patient. A large quantity

of urea is common to all diabetic cases, and especially to those
on animal diet. Regarding the cause of the high percentage
of urea in diabetes, the Rev. S. Haughton (a) remarks that it
comes from the excessive decomposition of protein substance
which takes place in this disease, quite independent of the
amount of bodily work done.

In a series of observations on two cases in University College
Hospital, Ringer(b) arrived at the following conclusions :—

1st. That after the influence of food on the urine has
entirely disappeared, a constant ratio is maintained between
the sugar and urea.

2nd. That after a purely non-amylaceous and non-saccharine
meal both the sugar and urea are increased, but that during
this increase the same ratio between them is observed. This
ratio being 1 of urea to 2·2 of sugar.

3rd. That under both these circumstances the sugar could
only be derived from the nitrogeneous elements of the body,
and, therefore, that some such ratio might, on à priori grounds,
have been expected.

Diabetes from Diminished Assimilation.

We shall now take examples of diabetes due to *diminished
assimilation*, and for this purpose select two cases from Hospital
practice, for the reports and analysis of which I am entirely
indebted to the combined labours of our clinical clerk, Mr.
Joseph Thompson, and my class assistant, Mr. J. S. Cluff.

One male and one female patient are again selected, and
that too of as near as possible parallel ages to the preceding
cases of the disease by excessive formation, so that the comparison of the results may be facilitated.

Fredk. F., aged 16, admitted into University College Hospital on October 15, 1864. Father and mother both living

---

(a) " Phenomena of Diabetes Mellitus." Dublin. 1861.
(b) "On the Relative Amount of Sugar and Urea in the Urine of
Diabetes Mellitus." *Trans. Med.-Chir. Soc.*, 1860, p. 323. *Year-book*, vol. II.

and healthy; family history generally good; no history of phthisis. Present illness commenced about a fortnight after last Easter (March 27), when patient first noticed that he made a great deal of water—six or seven pints daily—was very thirsty and drank much. His appetite was also very great; says "he used to eat anything that came before him." Has been treated dietetically, and, his father being a butcher, has had plenty of animal food. He improved for a time, but soon fell off again, and has been gradually getting worse; is much emaciated; used to weigh 105 lbs., but now only weighs 75 lbs.; skin very rough and dry; bowels regular; chest sounds healthy; no pain, cough, or night sweats.

The diagnosis of the case being of diabetes from mal-assimilation, the patient was placed on ordinary diet, including common bread, eggs, and a pint of beer; while to improve his general health a mixture with phosphoric acid and strychnine was ordered.

The patient's temperature was noted three times a day from October 19 to November 3, and found to average $97\frac{1}{2}°$ F., being three degrees below the ordinary standard.(a)

November 25.—Patient looks better. Says "he feels himself much improved; is not so thirsty; and makes less water." Goes home into the country to-morrow.

The following table shows the results of the analysis.

Date, 1864.			Quantity of Urine in cub. cent.	Reaction.	Sp. gr.	Quantity of of Sugar in grammes.	Urea.	Weight in pounds.
October 21	..	..	3480	Neutral.	1040	218·00	—	75
„ 26	..	..	5457	„	1037	286·10	—	77
„ 28	..	..	7000	Acid.	1033	280·00	—	77
„ 31	..	..	6200	„	1034	281·18	—	80

(a) Rosenstein was, I believe, the first to mention that in diabetes the temperature of the body is below the normal standard (*Virchow's Archiv.*) I find, however, that this is only invariably true in that form of the complaint arising from mal-assimilation.

Date, 1864.	Quantity of Urine in cub. cent.	Reaction.	Sp. gr.	Quantity of Sugar in grammes.	Urea.	Weight in pounds.
November 4	6508	Acid.	1037	260·8	—	78
„ 7	6660	„	1037	281·16	39·960	80
„ 9	6700	„	1036	341·6	40·200	80
„ 16	6840	„	1036	348·12	44·460	79
„ 18	5860	Neutral.	1034	322·17	41·020	79
„ 22	5790	„	1033	246·5	49·215	79
Received urine from the country :						
December 2	6300	—	—	187·29	—	84
„ 22	5735	Feebly acid.	1035	248·8	57·35	87
1865.						
January 26	5890	Acid.	1036	294·5	44·17	90
February 22	5270	„	1035	263·5	—	90
April 11	5580	„	1031	134·2	40·65	90

During the first few days after the patient's admission into
the Hospital the quantity of the urine was greatly diminished,
but as he became accustomed to the change, it again rose to
its average, and there remained until tinct. of cannabis Indica
along with four drops of laudanum three times a day was
administered (November 16), when it immediately fell 1000
cub. cent. in the twenty-four hours. The quantity of sugar
eliminated also decreased, but of course, in consequence of the
nature of the case, not to any marked degree. The patient,
since his return to Nottinghamshire, has still continued the
use of the remedy, the effects of which will be best appreciated
if I quote a passage or two from the last letter he wrote to Mr.
Joseph Thompson, dated April 10, 1865 :—" I have sent you
another sample of my water. . . I am a great deal better in
health and strength, and my weight is 90 lbs. without my
clothes. I am still taking Dr. Harley's medicine, and with
thanks," etc., etc.

The rapid improvement which took place in this case when
the diabetic system of treatment was discontinued, and the
patient was put back to ordinary food, is not a little remarkable. The improvement cannot be attributed to the change of

air or mode of life, for as we see it continued just the same when the lad had returned to his own home and former avocations. This is a most instructive case in more respects than one, for had the patient been treated by the ordinary routine system, the more sugar he passed, and, consequently, the worse he became, the more restricted would have been his diet; and I have little hesitation in saying that had such a line of treatment been pursued it is highly probable that ere now the case might have terminated fatally, for, as before said, at the time the lad came to the Hospital he was but a shadow of his former self, having lost 30 lbs. in a few months.

I have now to contrast a case of diabetes arising from malassimilation in a female with that from excessive formation in the lady's case previously cited. The nearest in age happens to be a very bad case—I may say a hopeless one, for the disease was much too far advanced before the girl came to the Hospital to render it probable that she will ultimately recover.

Sarah F., a milliner, aged 27, single, was admitted into University College Hospital on January 23, 1865. Has been an out-patient under Dr. Harley since December 9, 1864.

History.—Father died of "dropsy;" mother living (aged 72), has always had good health; has five brothers and sisters all healthy; no history of phthisis in the family. Patient states that until the last twelve months she has always enjoyed very good health; about this time (twelve months ago) she began to feel weak, and had a general feeling of *malaise;* she lost her appetite, and since last August (when she weighed about nine stone) has lost three stone weight. Three or four months ago, a few days after receiving a great mental shock, she first noticed that she made a great deal of water (five pints in twenty-four hours); she could eat nothing at this time, but was excessively thirsty, and used to drink large quantities of milk and toast water; lived chiefly on slops, arrowroot, &c.; took no solid food whatever; says "she has not moisture enough in her mouth to swallow." Her appetite since she has been an out-patient has materially improved. Still thirsty. Patient is tall, but very thin; skin dry, harsh, and rough; lips dry; flush on cheeks; no cough or night sweats; does not sleep well of a night, but has no pain; bowels very costive, will not act without medicine. Has not menstruated since last August, but up to that time her courses had come on too frequently—every fortnight or three weeks irregularly. Urine very light coloured; sp. gr. 1041; no albumen, but abundance of sugar.

Put on ordinary bread diet, including chop, beef-tea, milk, custard pudding, brandy ʒij.

Before coming under Dr. Harley's care the patient had been treated dietetically.

From December 12 until February 10, the patient took fifteen minims of tincture of cannabis Indica three times a day. She left the Hospital and became an out-patient again at her own desire.

Date.	Quantity.	Sp.gr.	Re-action.	Sugar.	Urea.	Weight.
	c. c.			grms. grs.	grms. grs.	lbs.
Dec. 10	3720	1037	Acid	186·00=2883 00	53·60=833·90	81½
,, 12	3642	1037	Neutral	214·20=3320·10	58·27=903·18	83
,, 16	3562	1036	Acid	178·25=2762·87	60·30=934·65	85½
,, 23	3720	1035	—	150·80=2337·40	—	—
Jan. 4	3825	1040	Neutral	161·45=2502·47	—	86
,, 7	2790	1038	Neutral	116·25=1801·87	—	—
,, 17	2945	1032	Acid	98·16=1521·48	55·95=867·22	—
Feb. 7	3038	1038	—	138·0 =2139·00	—	—
,, 10	3317	1034	—	144·50=2239·75	—	—
,, 18	2697	1039	—	134·10=2078·55	—	87
March 6	3100	1033	Acid	129·15=2001·82	—	—

Mr. Cluff informs me that in both this and the preceding case the analyses were always made on a sample of the forty-eight hours' urine, consequently the results are of double value. No one will, I think, venture to say any of these cases have been selected on account of their being favourable specimens for treatment, for in reality they were just the reverse.

Both of the two last cases had been considered perfectly hopeless by their respective Medical attendants, and yet we see that under a different line of treatment they both improved. The first one may ultimately get well; the last I am not sanguine about, notwithstanding that the strength of the patient has considerably improved.· When first brought to the Hospital, the girl was so weak as to require to be allowed a seat during her examination; whereas at her last visit she said she was able to walk to the Hospital, a distance of more than a mile; but even that distance seemed to knock her up. There are three bad features in her case:—Firstly, the want of appetite; secondly, the excessively costive bowels; and thirdly, the large amount of urea passed considering the small quantity of food taken.(a)　These four last-mentioned cases, I think, clearly illustrate the difference between diabetes resulting from *excessive formation*, and diabetes arising from *diminished assimilation*. I have only now once more to remark that as far as dieting is concerned, the treatment of the two classes of cases is diametrically opposite. The chance of success will therefore be in direct proportion to the correctness of diagnosis.

It ought always to be borne in mind that there are such things as *acute*, as well as, *intermittent diabetes.*· First, as regards the acute form of the disease.

Acute Diabetes.—Two remarkable examples of this kind are reported by Dr. Noble.(b)　One case was that of a boy aged 17, who died three days after the disease was discovered, and only a few weeks after he first felt ill. The other, a young lady, who died on the tenth day after the nature of her malady had been diagnosed.

(a) I have since heard from Mr. Clarke, of Gerrard-street, that this poor girl subsequently died while under his care, the most striking feature of her case being the impossibility of getting the bowels to act.

(b) *Brit. Med. Jour.*, Jan. 17, 1863.

Intermittent Diabetes.—In some patients, again, the peculiar condition of body inducing saccharine urine is not in constant operation, but comes and goes at irregular intervals. A well-marked case of this kind occurred in the practice of Dr. Ramsbotham, of Amwell-street, the particulars of which he has kindly furnished to me. They are as follows :—

" A lady, aged 72, applied to me in September, 1860, in consequence of sore tongue and intolerable thirst, which at that time had existed for several weeks. The peculiar smell of the breath and increased quantity of the urine (four or five pints each day), induced me to suspect diabetes. On examination, I found the specific gravity to be 1044, obtained a deep brown colour on boiling with caustic potash, and a copious red precipitate with sulphate of copper and potash. I gave mineral acids, with generous diet, omitting, as far as possible, all farinaceous food. By the end of a fortnight the specific gravity of the urine was under 1020, and without a trace of sugar. At the end of another fortnight the specific gravity became again 1044 and the urine loaded with sugar. This intermittent state continues up to the present period. The specimen of urine for the last twenty-four hours (November 26, 1861) has a specific gravity of 1018, and does not contain sugar."

Diabetes complicated with Other Disease.

Having already seen how saccharine urine is frequently the result of other affections, I have now to point out how other diseases frequently arise from the want of stamina induced by the diabetic diathesis. This is more particularly the case with that form of affection resulting from defective assimilation, in which, to use an insurance office expression, " it is a bad life." As I said before, Prout used to say of all diabetic patients, that they perpetually stood on the brink of a precipice. That

statement, though still perfectly true, has been fortunately
stripped of some of its terrors, since physiology has taught
us that, under judicious treatment, the life of the diabetic may
be made little inferior to that of patients labouring under
other chronic affections. Forewarned in this disease is truly
forearmed, so if the patient neglects the warning he has but
himself to blame for it.

Some few affections seem to be the direct effects of an excess
of sugar in the circulation, others the indirect result of the low
condition the excess induces, rendering the body unable to resist
the inroads of disease. Thus Mr. France has called attention(a)
to the frequency with which cataract is met with in the diabetic, and relates cases of its occurrence at an early age—30,
and even 19 years. The characters of these cataracts are
peculiar,—Firstly, they are symmetrically developed on both
sides; secondly, the lenses are increased in the antero-posterior
diameter, so as to interfere with the free play of the iris; thirdly,
the opacity attacks different strata of the lens at once; fourthly,
the colour and bulk of the lens denotes its soft consistency;
and, lastly, they do not come on till after the renal malady
has existed for some considerable time. Moreover, the eye
ought in no case to be interfered with, unless the patient is
completely blind, as these cases are not good for operation.

The cause of these cataracts has been well explained by Dr.
Richardson,(b) who was able to produce them artificially in
animals. They arise from a physical cause—osmose—a
tendency to equalise the density of the fluids in the lens with
those exterior to it,—in fact, the imbibition of sugar and the
transudation of liquid from the lens. Chloride of sodium and
other saline solutions produce the same effect. They are not,
indeed, true cataracts, and the proper treatment is to cure the

(a) *Dub. Hosp. Gaz.*, May 1, 1859, p. 136.
(b) *Med. Times and Gaz.*, March 31 and April 21, 1860.

diabetes, not to extract the lens, which will become quite right
of itself when the quantity of sugar in the circulation is reduced.

Cases of diabetes are occasionally complicated with albuminuria. In some instances the kidney disease supervenes in
the course of the diabetes. One such case has already been
mentioned; another occurred to myself. The diabetic patient, a
man forty years of age, was suddenly attacked with albuminuria
and dropsy after exposure to cold and wet. The albuminuria,
after a time, gradually disappeared; the dropsy likewise
diminished, but the saccharine state of the urine remained.
On the other hand saccharine urine occasionally makes its
appearance in the course of albuminuria. This I have frequently seen; but then, the diabetic condition was not only
trifling, but merely temporary, so that little importance
need be attached to it. It may, however, become permanent,
in which case, especially if the patient is aged, great fears for
his safety are to be entertained. I shall presently relate a
fatal case of this kind ; meanwhile, I may mention a most
interesting one which recently occurred in the practice of
Dr. Eastlake. The patient, a young gentleman aged 12 years,
had resided for a time at Calcutta. His urine, when I
examined it, was four ounces in quantity, had a specific
gravity of 1012, was highly albuminous, giving a copious
precipitate with heat and nitric acid, either separately or combined, and contained a considerable quantity of sugar. The
urine, when boiled with potash, became brown, and gave off
the odour of glucic and melassic acid. With sulphate of
copper and potash, on the other hand (it mattered not which
of these solutions was added first), a fine mauve, instead of a
blue, liquid was obtained. On boiling, the mauve, as is usual
in those cases, changed to a purple, then to a red, and lastly,
a considerable quantity of sugar being present, a precipitate
of the reduced oxide of copper was thrown down.

On questioning the patient and friends closely regarding

the origin of the disease, nothing more could be elicited than
that his legs began to swell about two years ago, and that on
his urine being tested, it was found to be albuminous, there
being not the slightest history of nephritis, either with or
without scarlet fever.

On carefully examining the abdomen the liver was found to
be enlarged (3½ inches perpendicular to nipple), and tender on
pressure. This led me to the idea that the albuminuria was
secondary to the hepatic affection. As regards the sugar, I
may mention that it does not seem to have been discovered
until the patient came under the care of Dr. Eastlake, which
was some weeks previous to our consultation.

The reason why I said that when sugar permanently appears
in the urine in the course of albuminuria fears for the safety
of the patient are to be entertained, is because it generally
indicates a loss of vital energy, which, sooner or later, leads
to a fatal termination. Dr. Quain has communicated to me
an interesting case of this kind. The sugar first appeared
three years after the albumen. The amount of the urine
passed shortly before the patient's death being ten pints, and
the specific gravity, in spite of the albumen, 1035.

It may be laid down as a rule, as I have elsewhere(a)
pointed out, that the supervention of permanently saccharine
urine, even when the amount of sugar is very small, in the
course of *any chronic disease*, is invariably to be regarded as a
most unfavourable symptom. It is not at all improbable that
cases of albuminuria and diabetes may be met with in the
human subject when these morbid affections are the simul-
taneous result of nerve lesion. The reason I think so is, that
I, like every other experimental physiologist who has worked
much at this subject, have occasionally induced both albuminuria and diabetes in puncturing the fourth ventricle of animals.

(a) "On Jaundice and Diseases of the Liver," p. 70

T

This accident has generally happened when the injury to the nerve substance has been made too high up.

A curious case of cancer of nearly all the viscera, conjoined with albuminuria of one kidney and saccharine urine of the other, has been reported by Dr. Gibb,(a) the history of which is as follows :—

A woman, aged 55, had been affected with cancer of the uterus for four years, and three years before two operations were performed for its removal. The disease, however, returned, and destroyed the neck of the uterus, the anterior wall of the vagina, and posterior part of the bladder, the urine dribbling from this general cavity for upwards of a year. She died on October 13, the left leg being in a state of gangrene. On examination after death, the lungs, pleura, and bronchial glands were found affected with cancer, the former containing distinct tuberculous masses as well, some as large as an egg. The liver weighed four pounds, and contained a number of circular cancerous tubercles. The spleen also contained a cancerous nodule. The right kidney weighed, with the fluid contained in its dilated pelvis, seven ounces and three-quarters; the left, much smaller, weighed, with its fluid, three ounces. From both extended dilated ureters filled with urine. The urine in the larger kidney was of the specific gravity of 1026, feebly acid, and contained sugar, as proved by the usual copper and other tests; that in the left was of the specific gravity of 1015, neutral, and contained much albumen. The condition of the urine had unfortunately not been previously ascertained, as none could be obtained from the bladder during the last nine months of the patient's life.

Diabetes may also be associated with chylous urine. From the observations of Dr. Babington,(b) indeed, it might almost

(a) Path. Soc. Trans., 1859.

(b) Todd's "Cyclopædia of Anatomy and Physiology:" article by Dr. Babington on the morbid conditions of the blood be concluded that a saccharine condition of the blood predisposes to this disease, for many of the patients suffering from diabetes whose blood Dr. Babington examined were found to be suffering from that condition called piarrhœmia (milky serum).

A few years ago I had the urine of one such patient sent to me for analysis by Dr. Charles Coote. The case, which was afterwards published by that gentleman(a), was more than usually interesting, in consequence of its origin having been traced to psychical causes (mental anxiety), and from its having run a very rapid course. The ordinary symptoms of diabetes, such as thirst, and excessive flow of urine, only existed for three weeks. The average amount of water daily passed during that time was from five to six pints, and of a specific gravity of 1031·34.

It has already been said that diabetes is often associated with two forms of inflammation. Boils and dangerous carbuncles are well known frequently to accompany it; and now it would appear from a paper in the *Union Médicale* of February 28, 1861, that even gangrene occasionally arises in the later stage of diabetes. This is not at all incomprehensible, seeing the great derangement to the process of interstitial nutrition to which a supersaturated saccharine state of the blood gives rise.

Lastly I have to call attention to impotence as a common accompaniment of diabetes, especially in men nearing or a little beyond the prime of life. Before me is the urine of a case of this kind now under treatment at the Hospital. The man J. W., aged 50, was sent to me a few months ago on account of diabetes. This case having been a very favourable one of defective assimilation, under a course of zinc as a tonic, the saccharine urine may be said to have entirely disappeared. At least, the sugar is now in such small quantity that for the

---

(a) *Lancet*, September 8, 1860.

last six weeks it has not been considered worth the trouble of
a quantitative analysis. Indeed, on the last occasion (two
months ago) that a quantitative analysis was made, Mr. Cluff
found it amount to only 1·2 grammes (18¼ grains) in the
twenty-four hours' urine. The patient is at present being
treated for the impotence which began shortly after he discovered that he was diabetic, which is now six months since.
During this period he has never had an erection, although he
has several times had emissions during his sleep.

Cases of this kind are not uncommon ; fortunately, however,
there is one good thing connected with them—namely, that
after the saccharine urine is stopped the steady employment
of a strong nervine tonic, such as strychnine, is generally sufficient to effect a cure.

Treatment.—Like all intractable maladies, diabetes has had
a great many specifics proposed for its treatment, and this is
not surprising, seeing that the number of specifics generally
increase in exact proportion to the irremediable character of
the disease they are proposed to cure. I am not, therefore,
going to take up time by even attempting an enumeration of
the names, far less of the vaunted virtues, of the specifics for
diabetes, but shall at once proceed to lay down a few of the
general principles upon which my own line of treatment is
founded.

Fortunately, medicine is of much use in this affection; and
although some cases, of course, are beyond all human aid, yet
many—I might almost venture to say the majority—are, to
some extent, under our control ; and although no one would
venture to say that he can remove the malady, yet in almost
every case we can substantially mitigate the more distressing
of the symptoms, and render the patient comparatively comfortable for the remainder of his life. Occasionally we have
the satisfaction not only of finding the symptoms improve
under treatment, but even the saccharine condition of the urine

entirely disappear. We must never, however, speak too sanguinely even of such a case, for a person who has once had diabetes is far from being in the same condition as one who never suffered from the disease ; for, just as with insanity, no matter however perfect the lucid interval may be, there ever exists the danger of a relapse, so with diabetes, however long the intermission may be, there always remains the danger of a return of the affection.

The two great types of diabetes, that due to *excessive forma-tion*, the other to *diminished assimilation* of saccharine matter, require, of course, as far as animal dieting is concerned, oppo-site modes of treatment; for while in the former class of cases it is a most important—I might almost say an essential— adjunct to the other treatment, in the latter it is either detri-mental, or, at best, of no use at all.

Even in the most favourable cases for restricted diet, we must never allow ourselves to be deluded into the idea that, because we are mitigating the symptoms, and reducing the amount of sugar in the urine, we are necessarily curing the disease, or we shall frequently be doomed to sad disappointment. In keeping a patient on restricted diet, we are merely with-holding from him the straw and mortar out of which the bricks are made,—not removing the makers—so that, as soon as the straw and mortar is refurnished to them, they will again be found at work as actively as ever. It is true that it occasionally happens during the withdrawal of the straw and mortar the makers dis-appear ; but this, unfortunately, is by no means invariably or even frequently the case; it is rather, indeed, the exception than the rule. We must therefore rely on other means for the removal of the makers. Of these other means I shall presently speak. Meanwhile, let me explain that by the term restricted diet we mean not only the avoidance of all sugars, and substances con-taining saccharine matter, but also of all kinds of food con-vertible during the process of digestion into sugar. The foods convertible into sugar in the digestive canal are those containing starch (not gums), such as arrowroot, tapioca, sago, flours of all the different kinds of cereals (wheat, barley, oats, peas, beans, etc.), potatoes, carrots, beetroot, parsnips, turnips, and other edible roots.

Green vegetables, on the other hand, such as spinach, cabbage, turnip tops, Brussels sprouts, and lettuce need not be forbidden, as they contain too small an amount of starch to do much injury.

As for animal foods, on the other hand, every imaginable kind of fish, flesh, and fowl may be indulged in, so that even on the most restricted diet the patient has still a large margin for selection—beef, mutton, pork, venison, poultry, game, and wild fowl, oysters, lobster, crabs, prawns, salmon, cod, turbot, etc., Iceland and Irish moss, calf's foot or gelatine jellies, butter sauces, and salad oils. The only true hardship, in fact, the patient suffers is the deprivation of ordinary bread, and that appears to be a more severe one than most people imagine. I have known patients in whom the craving became at last almost intolerable, as if nature were crying out for some indispensable element of food. In order to mitigate this hardship, a great number of plans of depriving bread of the forbidden element, starch, have been suggested, and many of them have been in a great measure successful. Thus, we have bran, gluten, almond, and glycerine breads and biscuits constantly kept in stock by many of our London bakers.(a)

After a time patients get very tired of these substitutes, so it is as well to know that we may occasionally indulge them with well done toast, or very crisp pulled bread, the extra

---

(a) Blatchley's, 362, Oxford-street, N. ; Donges, Gower-street North, W.C. ; Hill and Son, 61, Bishopsgate-street, E.C. ; Van Abbotts, 5, Princes-street, Cavendish-square, W. Persons in the country or any others who may desire to have the bread made at home may obtain pure gluten flour daily or weekly from Parsons, Fletcher, and Co., 22, Bread-street, E.C.

heat having destroyed a considerable portion of the starch normally contained in the article.

As regards drinks, all such as contain saccharine matter are to be avoided; such, for example, as sweet sparkling wines, whether they be champagnes, moselles, or hocks. An embargo is also to be put on all liqueurs and fruity wines, such as young port, Roussillon, etc.; sweet ales, stout, and porter are also to be shunned. If the patient is to be indulged in wines at all, let him have dry Lisbon, old Madeira, Manzanilla, or Amontillado sherries, Chablis, Niersteiner, or old Sauterne; sound clarets may also be indulged in. When stimulants are deemed requisite, brandy, whisky, rum, or Hollands may be used; but these ought always to be employed with caution for the reasons previously given, when speaking of the artificial production of diabetes by means of stimulants introduced into the portal circulation.

All that has now been said regarding regimen has of course only had reference to that form of diabetes arising from *excessive formation*. There are no restrictions either as regards food or drink requisite in cases springing from *defective assimilation*. On the contrary, the duty of the Practitioner is to select for his patient not only that which is most nourishing, but also that most easy of assimilation. He will often find, too, that such cases not only tolerate but even demand the free use of stimulants, in order to support the flagging vital energies, and enable the weakened organs to perform their work. I think if I were asked what is the best remedy for diabetes, I might venture to answer, in the language of Opie, when the student inquired what he mixed his colours with, "BRAINS, SIR." For to say that any one remedy or particular line of treatment is suitable to all cases of diabetes would be simply charlatanism of the worst sort.

For example, when diabetes arises from a traumatic lesion of the nervous system, it is not the symptom of saccharine urine that we treat. It, we know, will disappear when the nerve-lesion is healed. All our energies are, therefore, directed to the hastening of the healing process.

On the other hand, when the cause is not traumatic lesion, but disease of the nervous system, which we know is neither likely to disappear of itself nor by treatment—such, for example, as tumour of the pneumo-gastric nerve—our efforts are directed to the subduing of its effects.

A similar remark is equally applicable to those cases where the source of irritation does not exist in the nerves themselves, but in the organs which they supply, as, for example, in diseases of the lungs, uterus, or stomach. When the cure of these is beyond our skill, we have still to try and mitigate their effects by the administration of sedatives. A certain amount of discrimination is, however, requisite in the selection of the sedative. When the source of irritation is in some part of the nerve centre, which it is not advisable to narcotise, cannabis Indica is, for instance, to be preferred to opium. On the other hand, if it be deemed advisable to produce a sedative effect upon the brain, then opium would be the remedy selected. Again, should the seat of the disease be found to be situated in the pneumo-gastric nerve, conia is preferable to either of the preceding, as it possesses a special narcotic influence on that nerve.

In another set of cases, again, when the digestive organs appear to be at fault, hydrocyanic acid or other remedies of that class may be employed. In fact, as before said, we must suit the remedy to the special case, and when we fail in finding benefit from any particular remedy, we must try the effects of a combination. For it will be constantly found that a mixture of drugs do good, although each had individually failed to make any favourable impression on the disease.

As the majority of cases of diabetes are very chronic, we can easily afford to test the value of the treatment by testing the amount of sugar passed under each remedy, and change from one to the other until a beneficial result is obtained.

In cases of diabetes from *defective assimilation* we find the same diversity of treatment required. Some improve under nerve tonics, such as strychnine and phosphoric acid, others are benefited by preparations of iron, such as the sulphate, iodide, and citrate. Then again we find cases absolutely demanding the free use of both stimulants and sedatives. In fact, it is exceedingly interesting and instructive to glance over the various plans of treating diabetes which have proved successful in different hands. Thus we find that in one case M. Burquet found that from the first day of the administration of the proto-iodide of iron the thirst diminished, the sugar disappeared, and the strength returned.(a) Denican again declares that he has successfully treated diabetes by giving the patient equal parts of alum and extract of rhatany ;(b) while Semmola speaks with equal confidence regarding the benefits of electricity in this Protean affec-tion.(c) Day found the peroxide of hydrogen (ozonic ether) useful.(d) It was tried in University College Hospital by Sir W. Jenner and Drs. Reynolds and Fox. One patient died, and the other two did not improve.(e) No effect was produced on the quantity of sugar or water. Pavy also speaks unfavourably of it.(f) I have tried it in seven cases ; in two it acted beneficially. In one of them, a case of diabetes from malassimilation, ʒjss. of the peroxide t.i.d., reduced the quantity of urine from 80 to 50 ozs. ; the patient, aged 47, being on ordinary diet. In six weeks he was so much better as to be able to return to India, and resume his duties as judge. The bromide of potassium, as well as of iron, is exceedingly useful in cases of cerebral diabetes, and act

---

(a) *American Med. Jour.*, January, 1857. (b) *Gaz. Med. de Paris*, No. 31, 1861. (c) *Med. Times and Gaz.*, November 9, 1861. (d) *Lancet*, 12th January, 1868, and 20th March, 1869. (e) *Lancet*, 20th March, 1869. (f) *Lancet*, 13th March, 1869.

better still when they are combined with a little opium or hydrocyanic acid, as the requirements of the case may be. Codeine, Pavy says, is useful, but I consider its effects far inferior to those of crude opium. In one case, 2½ grs. had been taken, three times a day for nearly four months, and the patient (a lady aged 40) at the time she consulted me, was scarcely able to walk across the room. I at once put her on the bromide of potassium treatment, and she rallied sufficiently to be able to walk in her garden. The improvement was, however, of short duration, for, as I afterwards learned, she died at Margate where she had gone for change of air. The phosphatic salts of ammonia and alkalies, with limited restriction of diet, has been found useful by Basham.(a)   Chloral hydrate I tried in three cases during 1870, but without any decided benefit. Lastly, a purely skim milk diet has been strongly recommended by Donkin, but for remarks on this subject see page 345.

From all this we draw the important lesson that the disease is in no case to be treated by name, but according to its special requirements. In concluding these remarks, I must again insist upon the necessity of attending to the three following aphorisms :—

1. In treating cases arising from defective assimilation by restricted diet, although the amount of sugar eliminated may be reduced, we only hasten the termination it is our desire to avoid.

2. In no case ought we to trust to the specific gravity of the urine alone as an index to the amount of sugar passed, for it is an untruthful guide.

Lastly, we must always bear in mind that in cases of diabetes of the second class (*defective assimilation*) a decided and permanent improvement in the health of the patient is not inconsistent with a temporary rise in the elimination of sugar.

(a) *Lancet*, 10th April, 1869.

LECTURE XI.

CHEMISTRY.

As urine containing albumen may be either acid, neutral, or
alkaline, the most common, and, at the same time, the best,
method for detecting this substance is to employ heat and
nitric acid, at first separately, afterwards combined. In doing
so, a drachm of urine is put into a test-tube, and boiled.
Should the liquid thereby become turbid, albumen is most
probably present, but not necessarily so. To decide this
point, therefore, a few drops of nitric acid are added; and
should the turbidity caused by boiling be increased rather
than diminished on the addition of the acid, it is due to the
coagulation of albumen. If, on the other hand, the turbidity
either wholly or in part disappears, it has, in all probability,
been due to the precipitation of some earthy phosphates or
carbonates,—two inorganic substances which are often mistaken for albumen, in consequence of their being readily precipitated from neutral or semi-alkaline urine by heat. The
nitric acid re-dissolves both of these precipitates, the phosphatic without, the carbonic with, effervescence. Although
nitric acid precipitates albumen from urine without the aid of

heat, it must never be used alone; for when employed *per se*, it is a most fallacious test, in consequence of its giving an amorphous precipitate of uric acid in urine containing a great excess of urates, and a crystalline one of the nitrate of urea in urine loaded with that substance. The uric acid precipitate is frequently, the nitrate of urea precipitate rarely, mistaken for albumen. The application of heat in either case removes the difficulty by dissolving the precipitate, and enabling the substance to reassume its characteristic crystalline form during the cooling process.

In consequence of the above-mentioned source of fallacy in testing for albumen with nitric acid, some authors recommend the substitution of acetic acid; but it, too, has its drawbacks, for acetic acid throws down a precipitate of cystin when that substance exists in quantity in human urine; and if uric acid be in great excess, acetic acid is even sufficient to precipitate it. The microscopic examination of the deposit in all cases of doubt allows of its true nature being ascertained.

In testing human urine for albumen there is yet, unfortunately, another source of error—namely, the possibility of overlooking its existence when it is actually present. This may arise from two causes; one being the nature of the albumen, the other the condition of the urine.

In afterwards speaking of its physiology, it will be shown that one particular kind of albumen is a normal constituent of healthy urine, but as that variety of the substance is neither coagulable by heat, nor precipitable by nitric acid, we have nothing to do with it at present. What I now allude to, is the fact, that in the urine of disease albumen may exist, and yet escape detection, unless certain precautions be employed in order to ensure success.

In the first place, in alkaline urine a *small quantity* of albumen is not coagulable by heat; hence the necessity of always adding acid in order to avoid this source of error.

In the second place, some varieties of abnormal albumen are not coagulated by heat when only a little nitric acid is added, the addition of an excess being requisite to ensure the precipitation. On the other hand, another less common kind is occasionally met with where a great excess of acid re-dissolves the coagulum.

If, then, we wish to avoid these sources of error, when a discrepancy seems likely to arise, we must test repeated portions of the urine with different quantities of acid—say, with five, fifteen, and thirty drops of nitric acid to the drachm of urine. Thus, and thus only, are mistakes likely to be avoided.

There are still other reagents the employment of which in cases of doubt is attended with advantage. Such, for example, as absolute alcohol, chloroform, tannic acid, alum, acetate of lead, bichloride of mercury, and a variety of other solutions of the mineral salts—almost all of which precipitate albumen. None of these are, however, in experienced hands, at all to be compared, in point of accuracy, with heat and nitric acid.

The effect of the presence of sugar in causing albuminous urine to give a fine mauve colour with the sulphate of copper and potash has already been pointed out.

Specific Gravity.—It is generally stated that the specific gravity of albuminous urine is very low. But this is by no means invariably the case. In fact, it only holds good with one form of albuminuria, namely, that consequent upon structural change in the kidneys. A low specific gravity, therefore, aids in distinguishing that from the numerous other forms of albuminuria, in some of which I have met with it as high as 1032, or even 1035. These, however, were quite exceptional cases—the former being one of intermittent hæmaturia, the latter due to the co-existence of diabetes. In what is commonly called Bright's disease, the specific gravity ranges between 1005 and 1014. The usual average being 1010—12. In the other forms of albuminuria the specific gravity of the urine has no limit, it may be as low as 1005, or, as just said, as high as 1035.

The reason why the urine in certain cases of kidney disease has such a low specific gravity, is on account of the disorganised renal tubes being incapable of eliminating the urea and other urinary crystalloids. The lower, therefore, the specific gravity in these affections, the more dangerous is the case.

Quantitative Analysis.—In all cases of albuminuria a quantitative analysis is an invaluable aid to prognosis and treatment. It is best done by measuring off 50 c.c. (about two ounces) from the twenty-four, or, better still, from the forty-eight hours' urine of the patient, and adding it drop by drop to a couple of ounces of boiling distilled water acidulated with acetic acid. The capsule should be placed over the spirit-lamp, and its contents kept boiling vigorously the whole time the urine is being added. A drop or two of acetic acid should occasionally, during the process, be allowed to fall into the boiling mixture, in order to ensure its being always faintly acid. When the coagulation of the albumen is completed, the capsule is to be put aside, to allow of the coagula falling to the bottom of the vessel, and thereby facilitating the next stage of the process, which is the collection of the precipitate on a filter and thoroughly wash it with distilled water. After this, while still moist, the albumen is to be transferred to a watch-glass, dried, and weighed. The weight of the dried albumen represents, of course, the quantity in 50 c.c. (about two ounces) of the patient's urine.

The calculation for the twenty-four hours' urine is then simple enough. Suppose, for example, the quantity of urine passed be 1500 c.c., and the weight of dried albumen 0·2 grammes, then

$$\frac{0·2 \times 1500}{50} = 6$$ grammes of albumen in the twenty-four hours.

The quantity of albumen passed varies considerably in different cases, at different times, and in different forms of the disease. It may be from 1 to 2 grammes (15 to 30 grains) only, or it may amount to from 20 to 30 grammes (300 to 500 grains) in the twenty-four hours. In average cases the amount leaves about 10 grammes (150 grains) daily.

To follow the above method time and apparatus are alike requisite. By those, therefore, who have neither the one nor the other at their disposal, the following mode of procedure may be adopted; it may be called the "rough and ready method," and, although far inferior to the other, is yet much better than none at all :—Take the twenty-four hours' urine of the patient and dilute it till it measures 3000 c.c. (100 ounces).(a) Then, to two drachms of this diluted urine, in a test-tube, add ten drops of nitric acid, either after or before coagulation by boiling. When this is accomplished, place the tube aside until the precipitate is all deposited. The amount of coagulum yielded daily gives a rough relative approximation of the quantity of albumen passed, and by preserving the test-tubes, and comparing their contents from day to day, a tolerably fair idea of the progress of the case may be obtained.

In these cases, in consequence of the dilution of the urine, the coagulum seldom looks to be much. Having now not only ascertained that the urine of the patient is albuminous, but even the amount of coagulable matter passed, we are still but on the threshold of our diagnosis. The source and cause of the presence of the protein substance has yet to be discovered.

The coagulable matter found in the urine may have come directly from the serum of the blood, as in Bright's disease; it may be the result of a lesion of the kidney, as in renal calculus; it may be the product of inflammatory action, as in

---

(a) The amount of dilution is immaterial so long as it is always the same. Few patients pass more than 100 ounces of urine in the twenty-four hours; hence, I have made it the standard of comparison.

cystitis, or the secondary result of a variety of affections quite independent of renal disease. Many things have therefore to be taken into account ere we can arrive at a correct diagnosis even in a case of albuminuria. Before touching upon them, however, I must first say a few words regarding the physiology of our subject.

PHYSIOLOGY.

In studying albuminuria, it is well to bear in mind that the presence of albumen in the blood is a normal condition, while its existence in the urine is always abnormal. When I say the existence of albumen in the urine is abnormal, I mean of a kind and in a quantity sufficient to be detectable by heat and nitric acid. There is always albumen in healthy urine, but like the albumen found in the stomach, after being acted upon by the gastric juice, it is neither coagulable by heat, nor precipitable by nitric acid. I believe it, in fact, to be the effete albumen of our blood, excreted like urea or any of the other products of tissue metamorphosis.

Gigon was the first to call attention to the presence in healthy human urine of a substance coagulable by chloroform. He thought it was ordinary albumen, and that it had been previously overlooked in consequence of the tests not being sufficiently delicate for its detection. This, however, can scarcely be the case, seeing that we pass on an average 2·6 grammes (40·3 grains), in the twenty-four hours of this coagulable matter. Becquerel (a) says the substance described by Gigon(b) is simply mucus; but this cannot be the case, as Gigon finds that chloroform gives a precipitate when shaken with urine from which all the mucus has been removed. He has even shown that the same thing occurs in urine taken direct

---

(a) *Comptes Rendus,* Nov. 21, 1857.
(b) *Union Med. de Paris,* No. 12, 1858.

from the pelvis of the kidney, in which there is no mucus. Moreover, if the chloroform precipitate be collected, dried, and redissolved in acetic acid, it gives with ferrocyanide of potassium the reactions of albumen. In experimenting on this subject I found that absolute alcohol is even a better agent than chloroform, by which to extract this albumen from the urine, and by its means I was enabled to confirm Gigon's statement regarding the presence of albumen in urine devoid of mucus. The conclusion, however, which I arrived at was, as before said, that this albumen is not like that met with in the urine of disease, but like the albumen which after having undergone the modifying action of the gastric juice, is neither coagulable by heat nor nitric acid.

Experiments on animals have shown that unless the albumen of our food has been properly modified during the digestive process, it cannot be assimilated by the tissues, but acts in the blood like a foreign material, and is as such eliminated by the kidneys. Thus, for example, it has been found that when the albumen of the hen's egg is injected into the blood of dogs, it does not become incorporated with the tissues, but is rapidly thrown off along with the urine; whereas if the same albumen be put into the animal's stomach, and digested ere it enters the circulation, it is assimilated, and, consequently, does not appear in the urine in the shape of albumen coagulable by heat and nitric acid. This physiological fact gives us a clue to the well-known circumstance that temporary albuminuria often follows upon disordered stomachal digestion. Stomachal digestion being not only a dissolving, but a transforming process, we can readily understand how, when this transforming process is interrupted, either in consequence of the unsuitable quality of the food, or of some derangement of the digestive system, albumen may be absorbed into the circulation, not sufficiently changed to enable it to become incorporated with the frame,

U

and therefore, as in the case of the dog, appear in the urine.

There can be no doubt that the albumen found in albuminuria is not always in the same form. The mere fact, already alluded to when on the chemistry of the subject, of its occasional capricious behaviour towards heat and nitric acid is sufficient to prove this. There can be no doubt that even the urine itself has a powerful effect in modifying the reactions of albumen. Mr. Alfred H. Smee (a) has shown that while a current of oxygen gas passed through ordinary albumen, or albumen from the fluid of a spina bifida, at the temperature of the human body (98° F.) transforms a certain amount of it into fibrin, the same gas fails to produce any such result when passed through albuminous urine, even when the albumen is in so large a quantity as almost to solidify the entire urine when boiled.

Albuminuria may be produced by the physiologist artificially in animals by a variety of operative procedures. Section of the renal nerves in rabbits produces albuminuria (Michel). Division of the cerebral peduncles (Schiff) as well as section of the fifth pair of nerves within the cranium, is followed b a like result. Section of the spinal cord in the dorsal region of dogs I have found to be followed in the course of a few hours by albuminuria ; and all experimental physiologists are aware that puncturing the calamus scriptorius of rabbits, often times causes the urine to become highly albuminous. It has frequently happened to me also to bring on an albuminuria in animals, into whose portal circulation I had injected stimulants, for the purpose of rendering their urine saccharine.

In all these cases the albuminuria is most probably the result of the congestion of the kidneys, which almost invariably, to a greater or less degree, follows upon the above mentioned operative procedures.

_____

(a) *Proc.* Royal Society, June 16, 1864.

PATHOLOGY.

Since the introduction of the improved methods of chemical and microscopical investigation, the subject of albuminuria has undergone an entire revolution, and the time may now be said to have arrived when a philosophical arrangement of the different pathological conditions giving rise to it is not only desirable, but absolutely necessary. To grapple with preconceived notions, trample upon cherished theories, re-arrange data, and promulgate new principles, is, no doubt, a task not to be entered upon without serious consideration. Nevertheless, as we all know, if Medicine is ever to take her place among the exact sciences, some one of her votaries must gird on his armour, and boldly run the risk of obloquy in attacking old dogmas. While waiting the advent of a more able exponent, I shall, trusting to the support of the emancipated among my brethren, attempt to handle the subject of albuminuria in a manner similar to that in which I have elsewhere treated jaundice.

Unfortunately, however, at the very outset, there is an important difficulty to be overcome as regards the meaning which should be attached to a well-known title of Kidney disease. The united labours of Basham, Johnson, Christison, Virchow, Parkes, Bennett, Goodfellow, Begbie, Aitken, Quain, Stewart, Dickinson, Wilks, and others have shown that the term "Bright's disease" is at present indiscriminately applied to widely-differing pathological conditions, such as nephritis, cirrhosis, fatty and waxy degeneration of the kidneys, etc.,—renal affections which have no clinical connexion beyond what they derive from the associated symptoms of albuminuria and dropsy. If we desire to keep pace with the progressive advance of science, it is absolutely necessary for us either to limit the term "Bright's disease" to one particular form of renal affection, accompanied

with albuminuria and dropsy, or to abandon it altogether.
The latter alternative I am in no wise inclined to adopt, for
the name of the man who first discovered the connexion
of dropsy, albuminuria, and renal disease is, I opine, well
deserving of being handed down to all time. I therefore would
venture to propose that the name of Bright be perpetuated by
associating it with that form of disease which is not only the
one most frequently met with, but also that with which Dr.
Bright was most familiar, namely, the albuminuria of nephritis.
By so doing, we shall not only be enabled to make a philosophical arrangement of the different morbid states upon which
albuminuria depends, but at the same time render justice to
an honoured and deserving memory.

In studying albuminuria, there are two things that must
always be borne in mind : first, that there are certain forms of
albuminuria with dropsy, and certain forms without it ; and,
secondly, that all cases of permanent albuminuria are, in the
long run, always associated with an organic change of structure in the tissues of the kidneys, and that even in those cases
in which the renal organs were not primarily at fault, as, for
example, when the albuminuria springs originally from passive
portal congestion induced by heart or liver disease, paraplegia
or hemiplegia. Contrary to what some authors affirm, I assert
that permanent albuminuria is in all cases the *result*, and not
the *cause* of renal derangement, just in the same way as pus
is the *result* not the *cause* of an abscess. And just as an
abscess may be due to a great variety of causes, so may renal
derangement originate in a number of different morbid conditions—some constitutional, some local. When Johnson,
Walsh, Aitken, and other authorities, say that the textural
changes in the structure of the kidneys are only the local
expression of the constitutional disorder, as suppuration in a
gland may be but the local expression of constitutional struma,
I perfectly agree with them ; but when they go so far as to

say that the excretion of albumen by the kidneys is due to a
constitutional cause, and not to a ocal structural change, 1
beg to differ from them *in toto*. 'Tis true that in a limited
number of cases of albuminous urine the healthy kidney is
only eliminating a foreign material, for, as already shown in
the physiological part, when mal-digested albumen enters the
blood, the tissues of the body being unable to assimilate it,
the renal organs are forced to excrete this form of albumen as
they would any other noxious material; but these cases are
not only few in number, but have no similarity whatever with
the pathology of the ordinary forms of permanent albuminuria.
So small, indeed, is the proportion of cases of permanent albuminuria in which the primary cause cannot be directly traced
to the kidney itself, that they do not even, as a whole, amount,
I believe, to more than 20 per cent.

With these few remarks as a proviso, I may at once enter
upon my task, and as the great art in explaining a difficult
subject is to use as few and as simple words as possible, I shall
venture to put my views of the pathology of albuminuria, as
I did those of jaundice, into a tabular form.(a) It must be
borne in mind, however, that as all diagrammatic classifications
labour under the disadvantage of being more or less arbitrary,
it will be necessary, in order that the views of the pathology of
albuminuria here enunciated be thoroughly understood, that
the chemistry and physiology of the subject be previously
studied. Moreover, if the subjoined table is to be turned
to practical as well as theoretical account, the explanations to
be subsequently given of the different forms of albuminuria,
as met with at the bedside, must be carefully considered.

(a) Vide page 132 of the author on "The Pathology and Treatment
of Jaundice."

ALBUMEN IN URINE.

NORMAL (not detectable by heat and NO^5) $\begin{cases}\text{Always during health small quantity.}\\ \text{Detectable by absolute alcohol and chloroform.}\end{cases}$

ABNORMAL (Detectable by heat and NO^5).

Primary Albuminuria (cause originating in kidney).

- Temporary — Bright's disease $\begin{cases}\text{Traumatic,}\\ \text{Idiopathic, and}\\ \text{Scarlatinal nephritis.}\end{cases}$

- Permanent —
  - Fatty,
  - Waxy,
  - Chronic hypertrophy.
  - Cirrhosis.
  - Cancerous,
  - Tuberculous, and
  - Scrofulous degeneration.
  - Simple abscess.
  - Renal calculus.

- Digestive System $\begin{cases}\text{Improper food.}\\ \text{Imperfect digestion.}\end{cases}$

- Nervous System
  - Disease of Brain.
  - Congestion of Brain.
  - Softening of Brain.
  - Injury to Brain.
  - Hemiplegia.
  - Paraplegia.
  - Lesion of renal nerves.
  - Pregnancy.—Reflex nerve irritation.

Secondary Albuminuria (cause not originating in kidney).

Organic Disease (Passive renal congestion).
- Cardiac.
- Hepatic.
- Pancreatic.

Febrile.
- Small-pox,
- Ague,
- Typhus,
- Typhoid,
- Scarlet,
- Yellow,
- Puerperal,
- Rheumatic, and
- Remittent fevers.

General Affections (Active renal congestion).

Non-febrile.
- Pneumonia.
- Pleurisy.
- Bronchitis.
- Phthisis.
- Peritonitis.
- Gout.
- Scurvy.
- Erysipelas.
- Purpura.
- Cholera.
- Dysenteric Diarrhœa.
- Diphtheria.
- Constitutional Syphilis.
- Poisoning by Lead.
-     "   "   Arsenic.
-     "   "   Mercury.
-     "   "   Mineral Acids.

With the albumen present in the urine of health we have
now nothing to do, so I at once pass to the consideration of
the first great class, namely, that form of the affection in
which the cause of the albuminuria is to be found in the
kidney itself.

As indicated in the diagram, primary albuminuria naturally
divides itself into two well-marked varieties—a temporary and
a permanent variety—the former being by far the most com--
mon, the latter the most dangerous. Each of these will now
be specially and separately considered.

Bright's Disease.—The pathology of this temporary form of
primary albuminuria is simple enough, and may be summed
up in one word—namely, "nephritis;" for, be the exciting
cause what it may, the immediate cause of the albumen in the
urine is an engorged or an inflamed condition of the kidney.
Thus, for example, it may be the result of direct injury to the
loins, or arise from an attack of cold, either in a perfectly
healthy individual or in one rendered abnormally susceptible
to the affection by intemperance or other debilitating cause;
and lastly it may follow upon scarlatina, in consequence of
the kidneys having an excess of work precipitately thrown
upon them by a sudden check being given to cutaneous
exhalation. The diagnosis of the nephritic form of albuminuria, which is almost always acute and curable, is in general
easy; for, besides the history of the case, we have a most
important guide in the condition of the urine.

Should the nephritis be the result of injury, the case is clear
enough, for pain in the loins, the albuminous urine, and the
dropsy will be found to follow in its immediate wake. On
the other hand, when the disease is the result of cold and wet,
it is sometimes so insidious in its march as to give no indication of its presence until the existence of dropsy attracts the
attention of the patient. In the most common of this class of
cases, however, the general history is, that after exposure to

cold the patient suffers more or less malaise; is chilly and
feverish, occasionally even shivering ; has pain in the lumbar
region, with scanty and high-coloured urine ; next a puffiness
is noticed under the eyes, the feet begin to swell, and the
belly is observed to be enlarged. Should the attack be a severe
one, not only do the limbs become decidedly œdematous, but
the abdomen gets so filled with fluid that the patient is unable
to wear his ordinary clothes. All this may occur in a very
few days, and what is still more alarming to the patient and
his friends is, that the effusion does not always stop here, but
continues rapidly to extend until the whole trunk and upper
extremities are included in one general anasarca. Before
this time, however, Medical aid is generally sought, and the
progress of the dropsy thereby arrested. In men, I have occasionally observed that one of the most distressing results of
the effusion has been an excessive œdema of the genital organs,
the prepuce becoming, in some cases, so enlarged and distorted
as to obstruct micturition. At other times the liquid is chiefly
diffused into the cellular tissue of the scrotum, giving to it the
appearance of an immense scrotal hernia. In the spring of
1864 a little boy, aged 8 years, was brought to me at University College Hospital with the cellular tissue of his scrotum
so distended with fluid that it was nearly as large as his
head, the tense skin being at the same time so bright and
glistening, and the effusion so transparent, that most of the
students took it for an enormous hydrocele. As I pointed out
to them, however, the diagnosis between these cases and
hydrocele is very easy ; for in the one case the fluid is in the
subcutaneous cellular tissue only, and consequently the skin
retains the impress of the finger ; while in the other, the fluid
contained within the fibrous tunica vaginalis, prevents the retention by the tissues of the impress of pressure, however firmly it
may have been applied. It has been hinted that we frequently
meet with cases of Bright's diseases with very little or no pain,

or even malaise. I may now add that we occasionally
encounter others in which the œdema, especially at first, is so
trifling that it would fail to awaken alarm in the patient, were
it not accompanied with considerable discomfort, and scanty
and high-coloured urine. The smoky, red, or bloody-looking
urine at the same time generally yields a copious dark-
coloured deposit on standing, which further aids in arousing
the suspicions of the patient, which suspicions the Medical
attendant is able, by the aid of a little nitric acid and heat, at
once either to confirm or dispel. To diagnose the case properly, however, it is not alone sufficient to prove the existence
of albuminuria ; for that, as seen in the table, may occur
under a great variety of circumstances, and even when the
urine is so loaded with albumen as to allow of its being transformed into a solid mass—so solid that the tube may be in-
verted without danger of any falling out—we are still unable
to certify to the existence of renal disease until we have
called the microscope to our aid. If the case be one of Bright's
disease, in the urinary sediment will be found blood corpuscles,
epithelium scales, renal tube casts with their epithelial lining,
amorphous urates, and, perhaps, even crystals of uric acid or
oxalate of lime.

FIG. 1.

There are two points to which I must here direct special
attention, otherwise difficulties which need not exist may
come in the way of beginners. In the first place, it occasionally happens that a number of granular cells are found in

the field which look very like pus corpuscles (Fig. 2). They
are bigger than the red blood cells, and
smaller than the epithelium. These, in the
majority of instances, are simply white blood
corpuscles, and are most numerous after meals.
In some few cases, when the nephritis is accompanied by vesical catarrh, mucus cells are likewise abundant. In others, again (as in that of
a child aged 3 years, four weeks after scarlatina
and fourteen days after the first appearance of the dropsy
began) the white blood corpuscles are actually more numerous
than the red. This, however, is rare, for during fifteen years'
special observation I have only once met with it. The increase in the number of white corpuscles after food is common
enough, and is readily accounted for on physiological grounds,
it being now an admitted fact that after every meal a large
excess of white corpuscles are poured into the circulation.

Fig. 2.

The second source of difficulty to the beginner consists in
the fact that in some instances the blood corpuscles appear
nucleated. A remarkable case of
this kind fell under my notice in
June, 1859, the patient, a man
aged 32, being at the time under
the care of my colleague Professor
Hare. The albumen in the urine
had begun to diminish, the tube
casts and free epithelium to dis-
appear, when the red corpuscles were noticed to have the
appearance presented at Fig. 3.

Fig. 3.

Some of the corpuscles possessed one, others two perfectly
distinct maculæ. By slightly altering the focus of the lens a
halo became apparent round each macula, which then assumed
the characters of a black spot in the centre of a bright ring.
It was further observed that the spot was not in the interior

of the corpuscle, but only imbedded in its cell-wall, the appearance of its being in the centre of the corpuscle arising simply from the position in which the cells happened to be viewed. In many of the discs the spot was seen, as in the figure, to project beyond the cell-wall, thereby carrying, as it were, the latter with it. This irregularity in the form of the blood corpuscles I have seen still better marked in a case of hæmaturia. Some looked exactly like the developing nucleated blood corpuscles of the human embryo.

The peculiar appearances above described, which may prove so puzzling to the beginner, are due to the action of the saline urine upon perfectly normal blood cells. In 1860 Dr. Addison showed that sherry wine in certain proportions induced projections from the walls of blood discs, and in 1863 Dr. Roberts threw additional light upon the subject, by discovering that solutions of tannic acid and magenta bring into view a vesicle in the cell-wall of nucleated, as well as non-nucleated blood corpuscles. The inner vesicle being analogous, he thinks, to the primordial utricle of vegetable cells.(a)

Till now I have spoken of Bright's disease as it affects both kidneys simultaneously, but it occasionally, though much more rarely, happens that it only attacks one, and in this case there is albuminuria, and all the microscopic appearances already described, *but no dropsy*. Anasarca, in fact, only makes its appearance when not alone both, but the whole of both kidneys are so much affected as to be unable to excrete the urinary products. So long as one kidney remains healthy, these manage to escape from the circulation, and dropsy does not appear. This, indeed, is, I believe, the true explanation of the occurrence of those cases of Bright's disease without dropsy which have hitherto been considered so anomalous. Having diagnosed the case properly, the next thing is to watch its course, and try and guide it to a successful termination.

(a) *Proceedings* of the Royal Society, 1863.

The progress of the disease is best learned from the condition of the urine.

The first sign of improvement is a diminution in the number of the blood corpuscles ; the second, an increased elimination of urine, which is soon followed by a decrease of the dropsy. The quantity of urine daily voided after the first stage of the attack has passed is sometimes very large, being far beyond the normal standard. Occasionally, the twenty-four hours' urine of a patient in the second stage of Bright's disease will amount to something above 3000 c.c. (more than 100 ounces). And, what is still more important, is the fact that with this excessive elimination of water, there is often a corresponding excessive elimination of the organic and inorganic salts. Luckily for the patient it is so, it being this accumulation in the blood of these excrementitious products which in general is the immediate cause of a fatal termination. In one case, three weeks after the commencement of the attack, or, I should rather say, of the onset of the dropsy—for it was the first thing the patient observed—Mr. Alexander Bruce, my former assistant, found that the patient—a young woman, aged 29,—passed the following quantity and quality of urine :—

Water . . . 2525 c.c. = (81¼ ounces.)
Specific gravity . 1012
Urea . . . 25·25 grammes = 391·37 grains.
Uric acid . . . 0·233 ,, = 3·6 ,,
Phosphoric acid . 8·837 ,, = 137·0 ,,

And this, too, be it remembered, when the patient was taking very little food. As a case goes on improving the albumen diminishes, the specific gravity of the urine rises, and the tube-casts disappear.

Should, unfortunately, the affection resist the action of remedies, and run on to a fatal termination, which is rather the exception than the rule, the kidneys will be found enlarged, congested, sometimes softened, and always readily denuded of

their capsules ; but the capsule often has portions of the
cortical substance adhering to it. They may weigh in very
bad cases as much as twice the normal amount (which is about
10 ounces for the two). When examined microscopically,
small irregular punctiform extravasations will be found scattered among the urine tubes. As regards the urine tubes
themselves, some will be found entirely denuded of their
epithelial lining, others distended and blocked up with detached
epithelial cells, blood corpuscles, and amorphous granular
matter, giving to them a well-marked dark outline, readily
recognisable both in longitudinal and transverse sections.

The treatment of cases of Bright's disease will be considered
after the other varieties of albuminuria have been explained.
Meanwhile, it may only be remarked that the form of Bright's
disease here described is one of the most amenable to treatment of all renal affections.

We now pass to the consideration of the *permanent* variety
of *primary albuminuria*, which is, as shown in the table, met
with in the following morbid conditions of the kidney :—
Chronic hypertrophy, fatty, waxy, cirrhosis, tuberculous,
scrofulous, and cancerous degeneration.

The chronically enlarged kidney, which is by far the most
common accompaniment of permanent albuminuria, is in
general the result of repeated attacks of inflammatory action
(Bright's disease). It is likewise the ultimate product of
prolonged passive renal action, such as arises from cardiac and
hepatic disease. Besides this, however, all constitutional as
well as local affections (fevers, tumours, calculi, etc.) which
produce congestion of the kidneys may be said to pre-dispose
to their permanent hypertrophy.

The simply hypertrophied kidney is exceedingly difficult
to diagnose during life from the circumstance that so long as
the enlarged organ remains in a quiescent state—that is to
say, in an uncongested condition, or is not the seat of any

acute inflammatory attack—there are no symptoms sufficiently characteristic to admit of its ready detection, for the urinary secretion may be copious, normal in colour and specific gravity, as well as entirely free from any deposit ; so that unless some accidental circumstance directs the attention of the Physician to the kidneys, the existence of albuminuria may be readily overlooked. Even the œdema of the ankles is often in such cases so trifling as to escape the notice of both patient and Medical attendant, until some slight attack of renal congestion specially calls attention to the state of the kidneys and their secretion.

Regarding the morbid anatomical condition presented by the hypertrophied kidney, little need be said, for, except being much beyond the usual size, it looks to the naked eye exactly like a normal kidney. At least, this is the condition met with when the patient is suddenly cut off by some other disease. In other cases, again, the hypertrophied kidney is found engorged with blood, softened, and friable, in consequence of its being, or having recently been, the seat of an acute inflammatory attack. It is, indeed, surprising how prone the enlarged kidney appears to become congested, and even excessively so, without any very apparent cause.

When in a congested state it pours out large quantities of albumen, blood corpuscles, and epithelial tube casts. The urine at the same time becomes scanty and high coloured, the feet and legs rapidly increase in diameter, and the patient presents all the appearances of a well-marked case of Bright's disease.

Moreover, the hypertrophied kidney is frequently the precursor of both the fatty and the waxy forms of degeneration, and the almost invariable forerunner, I believe, of the cirrhosed state of the organ.

Fatty Degeneration.—Although the fatty form of kidney is frequently the result of chronic inflammation, it may occur quite independently of it, just as we find fatty degeneration arising from constitutional causes in other organs of the body.

Hence we often find that fatty kidney is the accompaniment of phthisis, of heart disease, of fatty liver, etc.

Its most characteristic symptoms are permanent albuminuria and dropsy, but as in the early stage of the attack it is impossible to say whether the albuminuria is to be persistent or not, these symptoms of themselves teach us but little. It is to the microscopic condition of the urine that we have to look for further information, and, fortunately, we need not look in vain, as a peculiar form of sediment not only reveals to us the nature of the pathological lesion, but even indicates the stage of the renal disorganisation. In the first place, the presence of oil globules in the tube-casts, and in the free epithelium, tells us that the albuminuria is the product of a kidney in a state of fatty degeneration; while, secondly, the amount of the oil globules and the quantity of albumen, together with the special character of the tube-casts present in the fluid, indicate the exact stage of the affection. Lastly, the supervention of an acute inflammatory attack upon a fatty kidney is at once made manifest by the existence of blood corpuscles in the urine, in addition to the presence of oily tube-casts and epithelium.

FIG. 4.

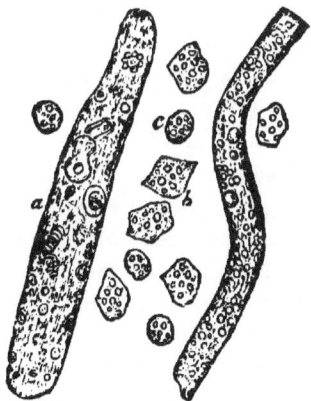

The accompanying drawing, which was kindly made for me by my former pupil, Mr. Ed. De Morgan, is a very faithful illustration of the condition of the tube-casts and epithelium met with in cases of fatty degeneration of the kidneys. The history of the case from which this drawing was taken is briefly as follows :—

A labourer, aged 36, was admitted into University College Hospital, under my care, on October 4, 1864. He acknowledged to have been a very intemperate man, but stated that he had, nevertheless, generally enjoyed good health. He could give no very definite account regarding the commencement of his illness, except that he noticed his legs first begin to swell about three months before his admission, from which time his health had gradually failed. There was no account of his having at any time had an acute attack of nephritis, nor of his present illness having been due to exposure to cold or wet. On admission, the legs and abdominal walls were exceedingly œdematous, but the face was only slightly puffy, and there was no anæmia to speak of. The urine was, however, highly albuminous, becoming almost solid on boiling. It was acid, contained oily tube-casts, and epithelium, as represented in the drawing; but there were also a few blood corpuscles and granular casts scattered among them. The case was, therefore, at once diagnosed fatty degeneration of the kidney, with the supervention of an acute inflammatory attack.

The patient continued much in the same state till November 9, when erysipelas, excited by a sore on the face, set in, and from then till December 5 he lingered on, sometimes better, sometimes worse, until he was suddenly seized with rigors and vomiting, followed by a worse outbreak of the erysipelas. He died on the 12th. The quantity of urine passed by this man varied from 70 to 116 ounces, and the specific gravity from 1010 to 1015.

On post-mortem examination made by one of my colleagues, the cortical substance of the kidneys was said to have appeared much increased in width, the line of demarcation between it and the pyramids being almost lost; while under the microscope the substance of the kidney was described as being in a state of granular and fatty degeneration, much free fat being scattered in globules over the field.

Where only one kidney is involved, fatty degeneration may run on to almost any extent, till at length the organ assumes

x

almost the appearance of a fatty tumour. One such specimen has at least been brought under my observation. The case occurred in the practice of Dr. Hullett Browne. The patient was a boy, aged 11, who had for a considerable time suffered from renal calculus. At the post-mortem examination, the left kidney was found to be 6½ by 3¾ inches in diameter, and so completely transformed into adipose tissue that only a mere trace of renal structure could be detected in the centre of the mass, faintly indicating, as it were, the outline of the original kidney. The right kidney was much enlarged, congested, soft, and friable. This was not, of course, an ordinary case of fatty renal disease ; but I quote it merely to show to what extent fatty degeneration of the kidney may proceed. If both kidneys are equally affected, death would, of course, inevitably ensue long before the degeneration could extend to anything like the stage here indicated. (a)

Between this and the healthy kidney there is every possible intervening grade of fatty degeneration. In the ordinary run of cases which terminate fatally, the kidney is found somewhat increased in size, greasy to the touch, and pale in colour, the cortical and medullary substances having generally much of the same appearance, and are so blended together that it is often extremely difficult to point out where the one begins and the other ends. A scraping of either part placed in a drop of water and viewed with the microscope presents, in an exaggerated degree, the appearances delineated in the figure ·-that is to say, the epithelium cells both in and out of the cubes are found loaded with oil globules. ·

Waxy Kidney.—Although this form of renal degeneration is not so common as the fatty variety, it is tolerably often observed, and is usually associated with a similar degeneration in some of the other organs, more especially of the liver

(a) The case is reported in the Path. Soc. *Trans.*, vol. xiii., p. 131. The specimen is in University College Museum ; No. 4720 of the catalogue.

and spleen. The waxy kidney is darker in colour than the
fatty, and is easily distinguished by its smooth texture and
glistening waxy aspect. To the touch it is firm, being sometimes indeed so hard as to take the impression of the nail
without tearing. On section, the medullary and cortical
substances are found to be blended together, and the mal-
pighian bodies in the latter distinctly marked as translucent
spots. The capsule is easily detached, and the surface beneath
is smooth, pale, and slightly mottled with small rosettes of
red vessels. The diagnosis of this form of renal disease during
life is not so easy as that of the fatty variety; still, however,
it can be accomplished by carefully examining the urine from
day to day, when the detection of what are termed hyaline
casts serve to indicate the pathological condition. The casts
are often absent for two or three days at a time; then for the
next two or three days they may be present. The annexed
figure, which was also made for me by Mr. De Morgan, indicates the appearance presented in the ordinary forms of hyaline
casts.

Fig. 5.

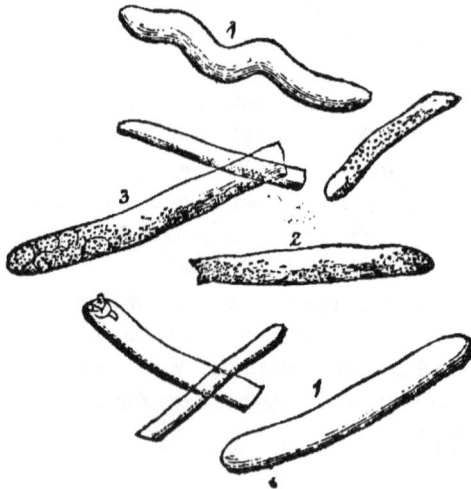

The patient who passed the casts from which the above drawing was made was a woman, aged 30, with a well-marked syphilitic history (secondary eruption, sore throat, loss of hair, tibial nodes, etc.). She attributed her urinary symptoms to an attack of cold, which induced headache, thirst, shivering, pain in loins, scanty and high-coloured urine, with dropsy. This occurred several months before her admission into the Hospital. At the time the above represented casts were found there was still albumen in the urine, but comparatively little dropsy, the lower limbs only being œdematous, the face having quite a natural appearance. As the waxy is sometimes associated with a partially fatty condition of kidney, it occasionally happens that both hyaline and oily casts are met with in the same patient's urine. This is always an unfavourable complication. Hyaline casts alone, though generally indicating an advanced stage of disease, are by no means invariably indicative of immediate danger. The waxy form of degeneration is sometimes almost entirely limited to one kidney, or portion of a kidney, and however severely affected the part may be, it causes little disturbance so long as the other kidney or portion of kidney, is healthy. With fatty kidney, on the other hand, it is usually the reverse. When once one part becomes affected the disease rapidly spreads; and this seems to be especially the case when the tissues of the organs have already become weakened by the pre-existence of other disease.

The waxy, like the fatty, kidney is prone to take on inflammatory action, and although during the quiescent stage of the disease no blood corpuscles are to be found in the urine, yet every slight attack of temporary renal congestion is apt to induce their appearance.

When examining the urine in a case of waxy kidney, a few epithelial cells are occasionally to be found scattered over the field of the microscope; but as all epithelial cells met with in the urine do not necessarily come from the diseased kidney, care must always be taken not to fall into an error of diagnosis, by mistaking urethral, bladder, or even vaginal epithelium, for renal. Renal epithelium, like the blood corpuscles, only appears in the urine of the waxy kidney, when an acute inflammatory attack or temporary congestion has disturbed its circulation. Grainger Stewart thinks that blood from the waxy kidney generally contains an excess of white corpuscles, but this, I imagine, can only occur when the spleen is also affected.

Cirrhosed Kidney.—The kidney, like the liver, after repeated attacks of inflammatory action is apt to become cirrhosed and atrophied. All the three varieties of enlarged kidney—the chronically enlarged, the fatty, and the waxy—are liable after a time to shrink and assume a granular appearance, but this is more especially the case with the former. The capsule at the same time becomes thickened, and so adherent to the renal tissue, that when it is being detached it brings away with it portions of the cortical substance, thereby giving to the surface of the organ a very roughened appearance. In acute nephritis, as before said, little patches of kidney substance often adhere to the capsule when it is torn off, but in that case the kidney is larger and softer, while in this case it is smaller and harder than natural. Besides, it has not the same granular appearance on section. The cirrhosed kidney sometimes even looks fibrous, in consequence of the obliteration of many of its tubes and blood-vessels. As the cirrhosed and atrophied kidney is also liable to attacks of inflammation, the appearances here described are not always those found after death, for the case often ends fatally during one of the inflammatory attacks, which of course considerably modifies the anatomical conditions.

The process which leads to atrophy of the kidney is thought to be very slow; some think it may extend over several years, but on this point I have not sufficient data at my disposal to

admit of my offering an opinion. The cirrhosed and atrophied
kidney is by no means easily diagnosed during life, for sometimes there is neither albumen nor tube-casts present in the
urine during several days at a time, and then it is only the
peculiar look, the pale, hydræmic face, the bleached lips, and
puffy eyelids, with diminished flow of urine, which lead us
to suspect the existence of renal disease. As a rule, the urine
is acid and scanty, both as regards the frequency of micturition and the amount passed on each occasion. When tube-
casts are detected they are usually small in size and somewhat granular in appearance. Their semi-granular nature has,
indeed, led me to the conclusion that their presence in the
urine in such cases, is perhaps not so much due to the atrophy
or cirrhosis of the organ, as to the existence of trifling temporary congestions, which though insignificant in themselves,
are nevertheless sufficient to disturb the balance of the circulation in an abnormally small organ, which even at the best of
times has more work thrown upon it than it can well perform.
The nature of the congestive granular tube-cast will be afterwards pointed out, when albuminuria arising from portal
congestion is considered. We must now pass on to albuminuria arising from tuberculous, scrofulous, and cancerous degeneration of the kidney, regarding which I have only a few
words to say at present, as their diagnosis and treatment
come much better under the head of bloody and purulent
urine.

In *tuberculous, scrofulous,* and *cancerous degenerations* of the
kidney, the urine is coagulable by heat and nitric acid, not so
much on account of the presence of ordinary, albumen as in
the preceding examples of disease, as from the existence in the
urine of organised albumenoid elements—such as tubercle and
pus cells, or blood and cancer corpuscles. In fact, in these
diseases of the kidney, although the urine is both coagulable
by heat and nitric acid, it is only when the other kinds of renal

disease, previously described, supervene in the course of tuberculous, scrofulous, and cancerous degeneration, that the coagulability of the urine dare be said to depend upon the existence of albuminuria, or that the affection ought to be classified in the category now under consideration.

Mr. Thomas Ballard has recorded(a) a very interesting case of encephaloid cancer of the kidney, in which there was the complication just described—namely, permanently albuminous urine of low specific gravity, associated with malignant disease of the kidney. These cases must, however, be rare. Before death, the lady, who lived to the age of 70, had frequent attacks of hæmaturia.

It ought never to be forgotten that ordinary renal inflammation may, and does under certain circumstances, run on to suppuration; but then, if the Medical attendant has carefully watched the progress of his patient by a diurnal microscopical and chemical examination of the urine, the case is easily enough distinguished from one of tuberculous or scrofulous degeneration,—for in the former case the appearance of pus corpuscles in the urine has been preceded by blood corpuscles renal tube-casts, and exudation cells; while in the latter the pus corpuscles have slowly and insidiously made their appearance independently of any inflammatory action. According to Basham, who is a good authority on such subjects, another of the peculiarities of nephritis ending in suppuration is the absence of dropsy. The existence of gravel or stone in an inflamed kidney is always a powerful provocative of suppuration but here, also, we have the previous symptoms of the case to guide us to a correct diagnosis.

Before quitting this subject, I should like to remark that the exact converse of what we have been describing occasionally occurs—namely, that a kidney, the seat of tuberculous,

(a) *Path. Soc. Trans.*, vol. x., p. 188.

scrofulous, or cancerous degeneration, may become suddenly the seat of nephritis, the supervention of which would be at once diagnosed by a change taking place in the urinary sediment, for, in addition to the elements of tubercle, cancer, or pus, the characteristic *débris* of Bright's disease, as already described, would become apparent.

SECONDARY ALBUMINURIA (CAUSE NOT PRIMARILY IN THE KIDNEY).

Albuminuria from Imperfect Digestion.—Every one who has had much experience with diseases of the urinary system is aware that cases of temporary albuminuria are occasionally met with in persons whose only symptoms are referable to disordered digestion. It may be that they are usually dyspeptic, or, what is still more frequently the case, they have been indulging in food that has disagreed with them. Thus, cheese has been known to cause albuminuria in children, and lobster or crab in grown-up people. The explanation of this is simple enough, as has already been explained in the physiological part of our subject, and which, therefore, need only be again alluded to here. In these cases there is nothing essentially wrong with the kidneys, and the albuminuria is simply due to the endosmotic equivalent of the albumen absorbed into the circulation being different from what it ought to be. Certain kinds of albumen, as before said, will not stay in the blood unless they have previously undergone modification by the digestive process. Hence, if the necessary modification be interrupted, either on account of the digestive organs being out of order, or by reason of the indigestible nature of the food itself, a temporary attack of albuminuria is the immediate consequence.

The effect of food on albuminurias of all kinds is much more important than most persons imagine ; for, as I shall afterwards show when on the subject of treatment, by diet alone we can at pleasure augment or diminish the amount of albumen eliminated with the urine.   Nothing further need therefore at present be said on the subject.

Albuminuria from Nerve Lesion or Reflex Irritation.—This, like the last form of the affection, is characterised by the absence of organic change in the structure of the kidneys.   But the analogy does not proceed further, for although the albuminuria is due to no structural change, it can, nevertheless, be clearly traced to a disturbance of the renal circulation.   It may be remembered that while on the physiology of the subject I mentioned that in the lower animals albuminuria can be produced at will by a variety of operative procedures—such, for example, as section of the renal nerves, division of the spinal cord, lesion of the cerebral peduncles, etc.   So we cannot be in the least surprised at the detection of albumen in the urine of patients labouring under similar nerve lesions, the result of disease.   As the cause of the albuminuria is apparent enough in the lower animals—namely, renal congestion—there need be, I think, but little doubt that in man the pathology of the morbid condition is precisely similar.   Cases of albuminuria in connexion with spinal paraplegia, hemiplegia, apoplexy of the base of the brain, etc., have been so frequently reported, that nothing further need be said upon the subject, but we may at once pass on to the consideration of that form which is least understood—viz., the albuminuria of pregnancy.

Albuminuria of Pregnancy.—Many years ago, when House-Surgeon to the Edinburgh Royal Maternity Hospital, I was led, at the suggestion of Professor Simpson, to examine the urine of every patient that entered the charity during three months.   The urine of those who came in after the pains of labour had already begun was immediately drawn off by the catheter, so that even in their cases all risk of error, from the accidental admixture of vaginal secretions, was avoided, and I was surprised to find how much less frequent the presence of albumen in th eurine of pregnancy was than I had been led to expect from the opinions expressed by previous observers. It did not amount to 4 per cent., which is actually less than a third the percentage of the cases of temporary albuminuria found to exist among female patients admitted into a general Hospital.(a)   As regards the frequency of puerperal convulsions, an almost similar conclusion was arrived at—namely, that they are not so exceedingly common as generally supposed, for although I have since seen them both in private and public practice, not one case of the kind occurred among the 450 women who were at that time delivered under my superintendence, and at least ten of whom were passing albumen in the urine. Moreover, even in those cases where the albuminuria is most severe, accompanied by dropsy, and apparently entirely due to the pregnancy, there may still be no convulsions, as the following case will show :—On May 30, 1851, a servant-girl, aged 20, was sent into the Edinburgh Infirmary by a Medical man who thought her case one of Bright's disease. Her urine was highly albuminous, her face puffy, her legs swollen, and her abdomen greatly distended.   On passing my hand over the abdomen I imagined that I felt an enlarged uterus, but on suggesting pregnancy, the girl indignantly denied its possibility.   It was true that she had seen nothing for the last nine or ten months, but to that she attached no importance, as she had often before been irregular.   She had never had morning sickness, never felt any movement, and besides, it was

---

(a) Among the women admitted into the Medical wards of University College Hospital, Parkes found 12·03 per cent. had temporary albuminuria (*Medical Times and Gazette,* January 1, 1859).   Drs. Elliott and Van Arsdale found only two cases of albuminuria in 112 pregnant women that they examined (*New York Journal of Medicine,* 1856).

impossible she could be pregnant. Facts, however, are stubborn things, and fœtal hearts being not usually heard in cases of simple ascites, I quietly, but firmly, told her my opinion that the albuminuria was associated with, if not the direct result of, pregnancy; at the same time adding that she might remain in the Hospital a few days while we tried to relieve the dropsy. In four days she was taken in labour, and within an hour after I was called to her bedside she was delivered of a full-grown healthy child, without a single convulsion or any head symptom whatever. From this time the albumen began to diminish in the urine, the dropsy to disappear, and on June 16, twelve days after delivery, not a trace of matter coagulable by heat and nitric acid was to be found in the urine. On July 14 she was dismissed in apparently perfect health.

I have now to relate two other cases of what is called the albuminuria of pregnancy, but which, as shall afterwards be explained, have an entirely different pathology. The following is an abstract of a case which was published in Germany in 1854.(a) The analyses were made conjointly by the late Dr. Gegenbaur and myself; the clinical history was furnished us by Scanzoni:—

The woman, aged 36, began to menstruate in her fifteenth year, and conceived for the first time in her twenty-fourth year, which conception, pregnancy, and delivery were all perfectly normal. Menstruation recommenced eight weeks after delivery, and continued at regular intervals till February 27, 1853, when she again became pregnant. The patient continued well till July 24, 1853, when she was attacked with ascites and anasarca of the upper and lower extremities, the outer genitals, coats of the abdomen, and of the face. Being

---

(a) Harley and Gegenbaur : Researches on the Blood and Urine of Albuminuria of Pregnancy. Bcit. zu Geburtskunde v. Scanzoni, Hf. II. 1854.

unable to work, she went to the Hospital in Schweinfurt, from whence she was discharged as cured in the course of six weeks. The patient attributed her illness to cold, which she said she had caught from being too lightly clad in damp, cold weather.

On September 18 she entered the Wurzburg Royal Lying-in Hospital, there to await her delivery. At the time of her entrance she was perfectly well, with the exception of an œdematous swelling of the lower extremities, which she ascribed to the long distance she had walked. After remaining fourteen days in the Institution, the patient again became dropsical, and the dropsy continued to increase so that on October 13 she presented the following appearance :—

Her pale face appeared swollen and puffy, particularly around the eyes; the upper extremities, especially the forearm and the hand on the left side, were very œdematous; the abdomen contained much fluid, and respiration was difficult, on account of the compression of the lungs; the heart's action was normal; the skin was extremely dry; the bowels torpid; and the scanty urine loaded with albumen. The fœtal heart was distinct. The patient was ordered a diuretic electuary, to remain quiet in bed, to take moderate nourishment, acidulated drinks, and the swollen extremities to be bound with a linen compress. During the next nine days both the œdema and ascites diminished, while the urine and perspirations increased. The amount of albumen passed was, however, much the same. From October 22, although the patient had ceased taking the medicine, the improvement still continued, and on November 5 2 ozs. of blood were withdrawn from the patient's arm, and the twenty-four hours' urine at the same time collected. The analyses of these fluids, as well as those subsequently made during the patient's illness, are given in the following table, in which, for the sake of more ready comparison, the constituents have been reckoned at so much per thousand :—

Blood.

In 1000 parts of blood were—

	1st analysis on Nov. 5.	2nd analysis on Nov. 17.	3rd analysis on Dec. 7.	4th analysis on Dec. 15.
Water . . .	825·96	829·404	789·220	768·782
Solid matter . .	174·04	170·596	210·780	231·218
Fibrin . . .	2·30	2·855	2·489	2·858
Albumen .	54·16	60·693	65·967	73·431
Blood corpuscles .	103·51	95·902	131·219	141·751
Extractive matter .	1·22	0·821	2·005	4·884
Inorganic salts .	12·78	10·315	9·100	8·287

In 1000 parts of serum were—

Water . . .	927·32	917·614	911·091	906·000
Solid matter . .	72·68	82·386	88·909	94·000
Albumen . .	60·69	67·269	76·154	86·538
Extractive matter .	2·91	3·752	4·505	2·462
Inorganic salts .	9·08	11·365	8·249	5·000

Urine.

In 1000 parts of urine were—

Water . . .	963·1	971·1	967·3
Solid matter . .	36·9	28·9	32·7
Inorganic salts .	22·9	15·9	17·5
Organic substances	12·88	12·8	15·3
Albumen . .	9·5	2·9	1·6
Urea . . .	3·39	5·43	10·9

The first analysis was made during the thirty-third week of pregnancy, at which time, although there was not much ascites, there was considerable œdema of the lower extremities. The qualitative analysis of the blood showed that it contained, besides the above-named ingredients, a large quantity of sugar and traces of carbonate of ammonia. The urine collected at the same time had a strongly alkaline reaction and nauseous smell—facts which led to the conclusion that the transformation of some of the urea into carbonate of ammonia was taking place in the circulation. On microscopical examination, the urine was found to contain a number of pale tube-casts, as well as numerous epithelium cells.

By the time the second analyses were made—twelve days later—the condition of the patient had improved, her complexion was more natural, and the dropsical swelling had considerably diminished. The blood no longer contained carbonate of ammonia, and only a small quantity of sugar was present. The urine had now become almost normal. The quantity had increased from 1169 cubic centimetres (38 oz.) to 1668 cubic centimetres (54 oz.), and the specific gravity had at the same time fallen from 1043 to 1025. Besides this, there were no longer any tube-casts to be detected. The improvement in the condition of the patient went steadily on until December 1, when she was safely delivered of a full-grown healthy child, without the occurrence of any nervous complication. On December 7—that is, seven days after delivery—the condition of the blood and urine, as shown by the third series of analyses, was still more favourable. The total quantity of urine then passed was 1792 cubic centimetres (58 oz.), the specific gravity 1015, and the reaction decidedly acid. Ten days after the confinement, the health of the mother was considered so good that she was sent as wet nurse to a resident family. Five days later, at the request of Scanzoni, she returned to the lying-in Hospital, and had over 2 oz. of blood abstracted from her arm, the result of which is given in the fourth table. Now, if we compare the first analysis of the blood with the figures given by Scherer of normal blood, the increase of the watery, together with the decrease of the solid constituents is very remarkable. The albumen is decidedly under the average of that found in non-pregnant females. The quantity of blood corpuscles is likewise less, whereas the fibrin and salts are more abundant than in normal blood. The quantity of solid matter in the serum is also much less, but the difference is here caused by the smaller quantity of albumen present, since the amount of salts approaches that in normal blood. The result of the second analysis corresponds with the improvement of the patient : there being *a decided increase of the albumen*, and decrease of the salts in the blood. The third analysis of the blood, seven days after delivery, shows, in comparison with the two former analyses, a considerable diminution of the water, with a corresponding increase of solid matters ; the fibrin is, however, less than in the second analysis. The great increase of albumen and blood corpuscles exactly correspond to the decrease of the albumen in the urine. In the fourth analysis, the water of the blood is seen to be still further diminished, and the solids at the same time as notably increased. The quantity of the salts is again less than in the previous analyses, so that all the analyses show a continuous decrease in these constituents.

The general conclusion to be drawn from the foregoing investigations is :—

1st. That in the albuminuria of Bright's disease there is less than the normal amount of albumen in the blood.

2nd. As the amount of the albumen in the urine increases, the quantity in the circulation proportionately diminishes ; and as the albuminuria decreases, the amount in the circulation gradually reapproaches the normal standard.

3rdly. The quantity of albumen and urea in the urine are in inverse proportion to each other—that is to say, where there is much albumen there is little urea, and where there is much urea there is but little albumen.

Lastly. The condition of the urine affords us a pretty correct idea of the probable condition of that of the blood.

I have now to point out an error into which people are very liable to fall—namely, to look upon the albuminuria of pregnancy, and pregnancy with albuminuria, as one and the same thing, while, in reality, they are perfectly distinct. When we speak of the albuminuria of pregnancy, we mean that the pregnant state induced the albuminuria, as was the case in the example first cited ; whereas when we speak of pregnancy with albuminuria we simply mean that a woman during the period of her pregnancy has been attacked with kidney disease, as in the case last cited.

I have now to call attention to a third form of albuminuria and pregnancy—namely, that in which a patient already the subject of kidney disease becomes pregnant. This is not only a graver complication than the "albuminuria of pregnancy," but, as a rule, even more dangerous, both to the life of mother and child, than a case of pregnancy complicated with kidney disease. The following case, of which I can only give a very brief abstract, will illustrate this remark :—

The subject was a lady, aged 30, who I saw in consultation with Dr. Magrath, of Teignmouth, and to whom I am indebted for the patient's history. Her mother died of diseased kidneys at the age of 45. The patient herself was healthy until she suffered from frequent attacks of intermittent fever whilst abroad. During her first pregnancy her legs were noticed to be œdematous, but, as albuminuria was not suspected, the urine remained unexamined. Her delivery was followed by a convulsion, for which she was bled to sixteen ounces. During the second pregnancy and confinement, nothing remarkable was observed, but after the third delivery she was again seized with convulsions, which were on this occasion arrested with chloroform. In the early part of 1863 she again conceived, and almost immediately before or afterwards, was, while recovering from a catarrhal fever, seized with hemiplegia. In the following month she was seen by Dr. Brown-Séquard, and, under the free administration of tonics, with change of air, she gradually regained the power of her arm and leg. In August of the same year, shortly before I saw her, her face and hands were for the first time observed to be œdematous. The urine was of a specific gravity of 1020, and was not noticed to contain albumen. About this time she suffered from headache, vomiting, with stertorous breathing, and muscular twitchings during sleep. At the end of August, when I examined the patient, the specific gravity of the urine was 1014, it contained a large quantity of albumen, coagulating to one-third of its volume, and with the microscope both blood corpuscles and renal tube-casts were detected. As the symptoms of uræmic poisoning continued to increase until the breath had a urinous odour, the induction of premature labour was had recourse to, during which she had a convulsion, which was arrested with chloroform. After this the patient made a slow recovery until October 24, when she was considered convalescent. The urine at this time was abundant, of a specific gravity of 1014, contained very little albumen, and only a few hyaline tube-casts.

On November 25 the quantity of urine was forty ounces, and the specific gravity 1021. Waxy tube-casts were still to be detected.(a)

In this case, as in the preceding, the union of pregnancy and kidney disease must be looked upon as accidental. In the former the albuminuria supervened during the course of the pregnancy; in the latter, the pregnancy in the course of the kidney disease,(b) each being quite independent of the other. The distinction to be drawn between these cases and

(a) In a letter I received on October 10, 1865, regarding another patient, Dr. Magrath incidentally alludes to the above case, and says Mrs. —— "keeps very well; the condition of the kidneys seems stationary."

(b) Although there is no data by which the exact period when the kidney disease commenced in this case can be determined, yet when we consider the history as a whole, it appears not improbable that the renal affection was first induced in the predisposed constitution by the repeated attacks of intermittent fever, from which the lady suffered at the commencement of her married life.

that first cited—namely, the albuminuria of pregnancy—is of the utmost importance both as regards prognosis and treatment; for while in the albuminuria of pregnancy it is to the pregnancy, and not to the kidneys, that we must look for relief, in the case of the kidney disease associated with pregnancy, as well as in that of the pregnancy associated with kidney disease, it is to the condition of the kidneys that we must specially direct our attention. The reason of this is easily understood when we remember that in the albuminuria of pregnancy the albumen rapidly disappears from the urine after delivery, whereas in the others the mere circumstance of delivery influences the albuminuria only in a secondary degree. The patient must be as carefully treated for diseased kidneys after the expulsion of the child as during its sojourn in the uterus.

As regards puerperal convulsions in cases of albuminuria, as far as my own experience goes—and I have nothing else to guide me—they are far more frequently the result of diseased kidneys than of the simple albuminuria of pregnancy. Convulsions, as we well know, frequently occur without pregnancy, and we cannot be surprised at their occurrence when pregnancy is superadded to the albuminuria. In the latter case the convulsive attack may be delayed until the time of delivery, or it may occur at any period of the pregnancy. There can be no doubt that albuminuria, from whatever cause, predisposes to epileptic convulsions, just as any other disease which impoverishes the blood, and thereby leads to malnutrition of the nervous system; for malnutrition of the nervous system is of all things the most fruitful source of epilepsy. In the second case cited we had ample proof of the impoverishing influence exerted upon the blood by albuminuria; and the absence of convulsions in it might probably be attributed to the circumstance that the blood had almost entirely regained its normal condition at the time of delivery.

That epileptic convulsions may occur in uncomplicated albuminuria in the predisposed I see no reason to doubt; and a well-marked case of this kind, which is briefly as follows, fell under my notice in 1851 :—

A dark-complexioned woman, of moderate development, who had been for nine years a strumpet,—she was then 28—was admitted on April 28 into one of the wards under the charge of Dr. Halliday Douglas. She had a dull, languid, and waxy appearance. Her urine was albuminous, of a specific gravity of 1008, and for several days after admission averaged 100 ounces per diem. She said her illness began with dyspeptic symptoms eight months before, and that shortly afterwards she had a fit, which had since recurred about once in every three or four weeks. On May 11—that is to say, thirteen days after her admission—she had a fit which lasted an hour, and from then until she died—seven days later—she had one, two, or more every day. She died immediately after coming out of one. From the 11th to the 18th she was in a state of stupor, and was not easily roused; but when awake she answered questions correctly. The urine fell from 100 to 50 ounces a-day during the last fortnight of her life. On post-mortem examination, the kidneys were found enlarged; one weighed six, the other five ounces; they were in a state of waxy degeneration. The woman had had syphilis. The capsule came off with difficulty; the texture of the organ was exceedingly firm, and of a pale colour. The separation between cortical and medullary substance was ill-defined. In the pyramids were several opaque points, which yielded a puriform fluid on pressure. When examined with the microscope, similar opaque points in the cortical substance were found to contain renal epithelium in all stages of degeneration, and numerous oleo-albuminoid granules.

Supposing this woman had chanced to become pregnant a month before she took ill, she would have been delivered just

about the time she died, and her death would in that case
have been most probably attributed to puerperal convulsions
from albuminuria, whereas, as we here see, the pregnancy
might in reality have had nothing at all to do with it. I do
not wish it for a moment to be supposed that I ignore the influence of pregnancy, and more especially of the effects of
delivery, in inducing convulsions in those otherwise predisposed to them.

All I desire is, to utter a strong protest against the oft-
repeated statement that puerperal convulsions are always the
result of the albuminuria of pregnancy, while in reality they
are much more frequently the concomitants of true kidney
disease, assisted by the effects of the puerperal condition, the
chief cause of the convulsions being the retention in the circulation of the excrementitious urinary products.

It is of vast importance to both mother and child to be able
to diagnose these cases ; for while in the albuminuria of pregnancy, unless the symptoms be severe, the case may almost be
left to nature, in pregnancy, associated with true kidney disease, energetic treatment is demanded, even, as we have seen,
to the induction of premature labour ; for even when the
pregnancy does not end in convulsions it never fails to act
prejudicially on the renal affection.

I would even go further, and say that pregnancy, in many
cases, is one of the exciting causes, not alone of albuminuria,
but of true kidney disease ; just in the same way as cardiac
and hepatic affections are—by keeping up renal congestion.

ALBUMINURIA THE RESULT OF ORGANIC DISEASE OF THE
HEART, LIVER, OR PANCREAS.

This is an exceedingly common class, and one demanding our
most careful consideration, as judicious treatment can here do
a great deal for the patient. The pathology of this class of
cases is very simple, mechanical obstruction inducing passive
renal congestion. As the albuminuria usually first makes its
appearance at a late period of the cardiac, hepatic, or pancreatic disease, it often happens that the patient succumbs to
the united effects of the double complication ere any well-
marked anatomical alteration in the structure of the kidneys
has had any time to occur. When such is the case, on post-
mortem examination the kidneys are merely found highly
congested. Should death, however, not have proved immediate, there is usually some structural change met with in the
renal organs, for, as we have already seen under a different
heading, the continuance of renal hyperæmia gradually, but
surely, leads to chronic hypertrophy, which in its turn ultimately induces a true structural change in the renal tissue.
In the class of cases now under consideration, the anatomical alterations met with in the kidneys, as well as the
functional derangements to which these changes give rise,
must be regarded as the secondary results of the diseased
liver, heart, or pancreas, as the case may be. But there
is also another set of cases where, although the double complication exists, the pathology is different. I allude to those
where the union of the renal and the other affection is, as it
were, accidental. That is to say, each disease has originated
independently of the other, certain causes having led to the
renal affection on the one hand, certain other causes having
induced the cardiac, hepatic, or pancreatic affection on the

other. The double complication in the latter class of cases
must be looked upon as a much more serious matter than in
the former.

For the sake of illustration, let us imagine the case of a
patient, who, after having for years been the subject of
chronic kidney disease, becomes suddenly attacked with
endocarditis, ultimately leading to permanent valvular obstruction, the tendency of this being, as we know, to lead
to hypertrophy of the heart. Here, then, is a case of a most
grave character ; for in addition to the cardiac hypertrophy we
have the increased tension in the arterial system produced by
the already existing kidney disease, also tending to the same
result. The evil does not even stop here ; for just as on the
one side the kidney disease tends to increase the cardiac
hypertrophy, so on the other the valvular obstruction lends its
helping hand to increase and perpetuate the kidney disease.

What holds good for cardiac affections holds equally good
for obstructions to the portal circulation induced by hepatic
or pancreatic disease. All these morbid conditions act and
react upon each other in such a manner as to render it exceedingly difficult to define how much of any given effect is due
to the one, how much to the other. Thus it often happens
that a patient who has been labouring under the combined
effects of cardiac and kidney disease suddenly becomes
seriously worse. The swelling of the feet increases, the pulse
becomes rapid and irregular, the breathing short and quick,
and within twenty-four hours after being in his usual
health he seems at the very gates of death. On examining
such a patient carefully, we most probably find a foul tongue,
a sallow complexion, an enlarged and engorged liver. Added
to this a bastard condition of lung—half pneumonic, half
bronchitic ; râles copious ; sputa sanguineolent ; dulness
irregular, the most distressing symptoms at the same time
being the inability to take a full breath and the impossibility

to procure sleep. No sooner does the poor patient lie down and shut his eyes than he starts up in a paroxysm of suffocation. Severe as these cases may be, they are by no means always without hope. I have seen patients as bad, and worse than here described, get comparatively well— at least, so well that they have lived for several months and been able to travel many miles. Perhaps this statement will be better appreciated if I relate a case of the kind; and the one I purpose selecting is that of a well-known publisher whom I attended with two other Medical men. When first seen, in conjunction with Dr. Walshe and Mr. Courtenay, the patient's condition was briefly as follows:—Great difficulty of breathing; respirations 38; feeble and irregular heart's action ; pulse 100 ; headache, sickness, and vomiting ; sputa pneumonic, red, rusty, and tenacious ; dulness over the lower pulmonary lobes on both sides of the chest ; loud mitral murmur ; œdema of the lower limbs ; scanty and highly albuminous urine, of a specific gravity of 1010. Coupled with these, there was an engorged liver and a sallow skin. In fact, a more hopeless-looking case could scarcely be imagined; for, although the patient could not lie down to sleep, he seemed gradually drifting into a state of uræmic coma. His breath and perspiration had already a urinous odour, and that, coupled with the uræmic pneumonia, made one regard the case with anything but agreeable forebodings. Strange to say, however, under very active treatment directed to the skin, lungs, liver, and heart alternately, this patient ultimately recovered, and was able to leave his bedroom within five weeks from the time that he was in the condition above described, and within a very few weeks more he was able to take a fifty miles' railway journey with perfect impunity. During the period of active treatment the case several times assumed so severe a character that both Dr. Walshe and Dr. Williams, who also saw the patient during his illness, agreed with Mr. Courtenay and myself in thinking that hours, not even days, would have limited his existence.

The influence of cardiac disease on renal affections is not yet all told, for it may lead to other complications than those already described—such, for example, as embolism of the kidney, of which the following case, kindly furnished to me by Dr. Bäumler, is a good illustration. It has the additional advantage, too, of being one in which the pathological condition was correctly diagnosed during life :—

E. K., aged 26, admitted into the German Hospital on October 3, 1864, with well-marked mitral disease; no dropsy ; rather difficult breathing ; crepitation at the base of the right lung.

October 4.—Urine scanty (about 700 c.c.), with deposit of urates ; contained only a trace of albumen, a good deal of pigment, and a considerable amount of urea, so that on putting strong nitric acid to it, the urine very soon crystallised.

5th, 6 p.m.—Sudden attack of severe pain in the right side below the twelfth rib, and in the right loin. Complete collapse ; extremities cold ; slight cyanosis ; no pulse could be felt ; sickness and vomiting, diarrhœa. Rallied after the administration of stimulants.

6th.—Pain and sickness continued.

7th.—Diarrhœa stopped. A little blood in the sputa.

From the 7th to the 12th of October the prominent symptoms were in the chest, the pain in the right side having then nearly subsided.

12th.—Feels better ; urine in quantity and appearance the same as on the 4th. *A little more albumen.* In the evening she was suddenly seized with vomiting, repeated purging, severe pain in the left hypochondrium, and collapse.

13th.—Pain continued very severe. (Great relief from subcutaneous injection of ¼ gr. of morphia.) The pain was greatest just below the twelfth rib, and radiated to the left loin.

15th.—No more diarrhœa. Urine 500 c.c.

16th.—Urine 1000 c.c., contained *more* albumen than on the 12th. The field of the microscope was covered with tube-casts, transparent and granular, and blood corpuscles.

22nd.—The same amount of albumen; felt better; œdema of the legs.

23rd.—Chest symptoms increased; considerable effusion into the right cavity of the pleura.

25th.—Less urine, less albumen.

27th.—Beginning jaundice.

31st.—No albumen nor casts.

November 2.—Urine about 1000 c.c., without deposit; large amount of bile pigment; *no* albumen.

11th.—*Not a trace of albumen* in the urine. Died on the same day.

The post-mortem examination showed a small fibrinous deposit in the right kidney in a state of shrinking; the upper part of the left kidney entirely diseased; one of the branches of the renal artery plugged up; the plug was about one inch long.

ALBUMINURIA THE RESULT OF GENERAL AFFECTIONS.

This includes a large group of diseases, which with advantage may be subdivided into the two varieties of febrile and non-febrile affections. Among the diseases composing these groups may be mentioned—

1st. Small-pox, typhus, typhoid, scarlet, yellow, rheumatic, puerperal, and remittent fevers.

2nd. Pneumonia, pleurisy, bronchitis, phthisis, peritonitis, gout, scurvy, erysipelas, purpura, and cholera.

The urine in all these cases may be coagulable under three different conditions. In the first place, the kidneys may only permit the serum of the blood to escape ; in the second place, not alone may the unorganised serum escape, but even the organised blood corpuscles, thereby giving to the case somewhat the appearance of acute Bright's disease, the urine being both coagulable and smoky, or even red. Thirdly, and lastly, in the severer forms of disease, especially in those assuming a malignant type, the walls of the blood corpuscles are broken down, and their disorganised contents—hæmato-globulin—escape, giving to the urine the colour and chemical reactions, without, however, the microscopical characters, of hæmaturia. Such a state of things is occasionally seen in putrid typhus, malignant scarlet fever, and severe purpura. I shall afterwards have to relate cases, under the head of intermittent hæmaturia, which closely resemble these, in so far as the urine contains disorganised blood constituents without blood corpuscles.

As regards the pathology of the albuminuria of general affections, all that need be said is that it is essentially due to renal congestion, and that there is usually but little difficulty in diagnosing the case, from the circumstance of its coming and going with the other affection. It is true that active congestion may also supervene in the course of one or other of the forms of primary albuminuria, but even then we have it in our power to recognise its existence by a peculiar form of renal tube-cast—the granular tube-cast. Granular casts were at one time thought to be characteristic of contracted kidney, but this I have long since proved to be an error. They may occur in cases of contracted kidney, as they may occur in any form of albuminuria ; but the cause of their presence is then, as it is now, active congestion.

Some idea of the appearances presented by granular tube-casts can be obtained from the accompanying illustration. The casts may be large or small, quite full, or only partially filled with granular matter, the size of the cast depends

on the part of the tube in which it was formed; the amount
of granular matter it contains on the condition of the kidney.

FIG. 6.

Quantities of escaped granular matter and epithelial cells may
also be irregularly scattered over the field of the microscope.

Before passing to the subject of treatment, I would again
call special attention to the fact that in every form of chronic
kidney disease, the organ is liable—more liable than the healthy
kidney—to occasional attacks of acute congestion or of inflammation, and that under such circumstances the appearances
met with in the urine become modified and complicated
accordingly. Fatty casts are found mingled with blood corpuscles, waxy casts mixed up with pus cells, granular casts
surrounded with oil globules, and so on, thereby throwing
almost insurmountable difficulties in the path of the beginner.
Care and experience will, however, soon clear the way
and show that the apparent confusion is but perfect order
not properly understood.

TREATMENT.

At no very distant date the treatment of kidney diseases was regarded as an almost hopeless task. As our knowledge of their pathology and our powers of diagnosis have, however, advanced, we have gradually gained confidence in the value of remedies in such affections, and we might almost say that it is in the power of the enlightened Practitioner not only to check the milder forms, but even to control and mitigate the symptoms arising from the severer renal affections. To be successful in treatment we must be correct in diagnosis, and taking for granted that the nature of the case has been thoroughly made out, the general principles upon which we are to found our treatment are as follows :— The first and great principle of all is to give rest to the diseased organ, for just as the function of the diseased eye is to be kept in abeyance while nature is removing the lesion, so must the diseased kidneys have rest from their function in order that their healthy condition may be restored. We cannot, unfortunately, lay up the kidneys as we can a fractured limb, nor entirely arrest their function, as we can that of a diseased eye. Nevertheless, we can reduce their labour to a minimum by making other organs for a time perform an essential part of their office.

There are three channels by which we can draw off from the system the excrementitious materials which normally fall to the lot of the kidneys to eliminate—namely, the bowels, skin, and lungs. By keeping up an excessive action in these, we can not only remove for a time a portion of the burden from the over-taxed kidneys, but in those cases where dropsy has already become a distressing symptom give great relief to the patient.

The value, indeed, of pressing into our service the vicarious action of the bowels, skin, and lungs in the treatment of renal disease it is scarcely possible to overrate, for we can not only by such means give rest to the kidneys and diminish dropsy,·but even mitigate the more serious symptoms which are ordinarily included under the term uræmic intoxication,(a) but which, more properly speaking, are the direct result of the combined effects of the retention in the circulation of all the excrementitious products, organic as well as inorganic, which normally fall to the lot of the kidneys to excrete. The skin and lungs are more powerful auxiliaries in the elimination of urinary products than is generally supposed, for, as we have shown in our first lecture, the cutaneous perspiration does not only carry off water, but many of the organic as well as inorganic urinary salts. Thus it has been found that the sweat even in health contains urea, uric acid, phosphates, and chlorides, while in disease, in addition to these, it contains many abnormal compounds, uch even as the insoluble oxalate of lime. Pulmonary exhalation, too, as was then shown, may be almost of equal service, for in the expired air of even healthy men have been detected urea, uric acid, urate of soda, and urate of ammonia. (b)

The vicarious action of the bowels is to be induced by the internal administration of mild or drastic purgatives, according to the constitution and condition of the patient. When there is much dropsy, elaterium is a favourite form of purgative, but in cases of kidney disease it is usually advisable

---

(a) The symptoms of uræmia have already been given under the head of Urea, p. 53.

(b) " In the *Deutsch. Arch. f. Klin. Med.*, vol. vii., parts 5 and 6, Dr. Deininger published the case of a boy five years of age, who suffered from anuria renalis for a whole week, and on whose skin urea was discovered. Five cases of this kind have been recorded by Jürgensen and Leube ; but the issue was fatal in all, whereas Dr. Deininger's patient recovered. This vicarious function of the skin is therefore proved, and should encourage us to use diaphoretics whenever the action of the kidneys is diminished, or altogether abolished."—*Lancet*, 17th Dec., 1870.

to administer it along with hyoscyamus, as it not unfrequently brings on an exhausting diarrhœa, especially if given after the symptoms of uræmic poisoning have already set in.

The vicarious action of the skin may be induced either by .the internal administration of diaphoretics or the external use of the warm bath, vapour-bath, or hot-air bath. The two latter not only increase the cutaneous, but also augment the pulmonary elimination of urinary products. This is especially the case with the hot-air bath.

One of the easiest of these in application is " Allen's port-able Turkish bath." It can be used by a patient lying in bed, as here represented, (Fig. 7), or what is still more convenient

FIG. 7.

(when the patient is strong enough), while in a sitting position, Thus, for example, the bath is put (with its disperser on) under a kitchen chair or stool, on which should be placed two or three folds of flannel. The whole person, except the head, is then packed in a blanket, having the ends well on the ground to prevent escape of heat. The feet being placed on a footstool, with a pan of hot water or not, according to inclination, and the bath may be taken in this way from fifteen to forty minutes. After which, clear the skin with soap and warm water, then sponge with cold water, and rub dry briskly with rough towels. The towels can be warmed ready for use, by placing them on the back of the chair while the bath is being taken. Even the calomel vapour bath may occasionally be used with advantage, in cases of albuminuria supposed to have a syphilitic origin. But before resorting to this line of treatment, one must be perfectly sure that the diagnosis is correct, for the strength of patients with albumen in their urine cannot be lowered with impunity. It is much easier to take away than to give strength to them.

I may here remark that the usual practice of trying to diminish rather than to increase the urinous odour of the sweat and breath in cases of advanced kidney disease is greatly to be reprehended ; for instead of trying to check, we ought, on the contrary, to assist Nature in her laudable efforts to rid the circulation of the deleterious agents that are gradually extinguishing the life of the patient.

Regarding the action of diaphoretics, which, as is well known, are exceedingly useful in chronic kidney disease, a word or two may be added. The most common form is Dover's powder, and from this containing opium, some have even ventured to administer that drug in combination with other substances than ipecacuanha. I must here, however, call attention to the fact that, although Dover's powder may be given with impunity, opium can seldom be employed in kidney affections in any other form without a certain amount of risk. More than one example of its deleterious effects in such cases has come under my notice, and in 1861 (a) a fatal case actually occurred in one of our London hospitals, where a man, aged 45, labouring under kidney disease, died after having taken only one grain of the acetate of morphia.

In cases where there is a tendency to convulsions, even Dover's powder must be cautiously used. The cases best suited for the administration of Dover's powder are those of renal disease complicated with dropsy. Ten grains given at

---

(a) *Lancet*, June 8, 1861, p. 575.

bedtime will often be found to relieve the patient, not only by increasing the perspiration, but also by augmenting the flow of urine.(a)

The object of giving diaphoretics in chronic kidney disease is, of course, to keep up cutaneous perspiration. For a similar reason, patients ought always to wear warm clothing; and this rule ought never to be departed from, even in cases where the patient is comparatively well, so well that complete recovery may be said to have taken place. In the latter class of cases, indeed, every precaution against the effects of cold ought to be had recourse to, for a patient who has once had an attack of kidney disease is ever afterwards much more liable to another attack than one who has never been so affected.

As regards the employment of diuretics in the treatment of kidney disease, a few words are here necessary. In the first place, it ought never to be forgotten that in acute Bright's disease, as well as in the first stage of all inflammatory and congestive attacks occurring in the course of chronic kidney affections, diuretics are inadmissible. In the second place, it must be borne in mind that great care should always be observed in their selection; for a diuretic which will prove beneficial in one form and at one particular stage of renal disease will often not only do no good, but actual harm, when administered in another form or at another stage of the same attack. Thus, whenever the albuminuria is the result of active congestion, the antiphlogistic variety of diuretic—such for example, as a combination of bitartrate, acetate, or nitrate of potash and digitalis—is to be selected; whereas in the absence of active congestion, and more especially when the vital powers of the patient are low, the stimulating variety of diuretic may not only be used with impunity, but with actual advantage; such, for example, as squills, buchu, uva ursi,

(a) Some judicious remarks on this subject will be found in the *Med. Press* of June 8th, 1870, by Dr. Macdonald.

belladonna, succus scoparii, spiritus juniperi, or spiritus
ætheris nitrosi. The reason why the employment of diuretics
often does harm in acute kidney affections is readily understood when we recollect that they have always the tendency
rather to increase than diminish the flow of blood to the
already engorged organ.

It ought never to be forgotten that the most potent diuretics
often fail to produce diuresis until vicarious action of the
bowels or skin has been established; whereas, after a smart
purge or a profuse perspiration, a comparatively small dose of
a diuretic, even a draught of cold water or a cup of weak tea,
will induce an increased flow of urine.

In the next place, the employment of antiphlogistics is in
many of the inflammatory forms of kidney disease of essentia
service. The most powerful of these is, of course, the local
abstraction of blood, either by leeches or the cupping-glasses.
But even general depletion by the lancet may, in some acute
cases, be had recourse to. This remark is specially applicable
to puerperal albuminuria, accompanied with convulsions.

In former days, indeed, bleeding was regarded as the
sheet-anchor in puerperal convulsions. It is only within the
last twenty-five years that the use of the lancet has died out :
in this, as in most other affections, I might have said
unfortunately died out, for I believe that venesection may
sometimes be had recourse to with great advantage.

In 1850, while house surgeon to the Royal Maternity
Hospital, Edinburgh, I had the opportunity on more than one
occasion of witnessing the benefit of free depletion, and
although I cannot venture to go so far as Professor Dyce, of
Aberdeen,(a) who recommended the fearless use of the lancet,
and other lowering remedies in this alarming disease, no
matter what be the constitution of the patient, the duration of
the pregnancy, or the stage of the labour, I think a return

_____

(a) See the *British Med. Journal* of 15th April, 1868.

to the use of the lancet, in those cases where the convulsions are apparently due to the engorgement of the cerebral vessels, highly desirable. Its indiscriminate use, however, I cannot but regard as highly unphilosophical with our present knowledge of the pathology of convulsions. (a) Just as in many cases of inflamed lung the general condition of the patient prevents the employment of active means, so also in the case of the inflamed kidney we must occasionally abandon the antiphlogistic line of treatment, and content ourselves by merely diverting for a time the course of the circulation, either by the application of dry-cupping glasses, counter-irritation, or of hot fomentations to the loins.

I ought not to omit to mention that the unloading of the portal circulation by a smart calomel purge will often prove an important adjunct to such measures.

ANÆSTHETICS IN PUERPERAL ALBUMINURIA WITH CONVULSIONS.

When in the course of puerperal albuminuria convulsions have once manifested themselves, one of the readiest and perhaps the safest means of subduing them appears to be by the free administration of an anæsthetic. Chloroform has been most commonly employed for this purpose, but as insensibility has usually to be kept up for a considerable time, it is much safer to employ the mixture which I proposed some years ago, and which received the approval of the Chloroform Committee of the Royal Medical-Chirurgical Society.(b) The mixture consists of one part by measure of pure alcohol, two parts of chloroform, and three parts of sulphuric ether. A ready way of remembering the composition of my mixture is rhyming the first letter of each of the ingredients, Alcohol, Chloroform, Ether, with their quantities employed. Thus it rhymes :

$$A. \ C. \ E. = 1 \ 2 \ 3.$$

(a) See remarks at p. 319.   (b) Trans. Roy. Med. Chir. Soc., vol. xlvii. p. 339.

I have no hesitation in saying that this mixture forms the
safest and most agreeable anæsthetic hitherto proposed. If
medical men were more intimately acquainted with the
powerful chemical changes which chloroform exerts on the
constituents of the blood, even when taken into the system by
the lungs, they would probably employ it with more reserve
than they do at present.(a)

The advantages of the alcohol, chloroform, and ether
mixture need not be here further alluded to, as they were
fully dwelt upon in the report of the before-mentioned
committee. All I shall say is, that it has special advantages
in cases of puerperal convulsions, in which a state of
anæsthesia requires to be kept up for some hours.

Hydrate of chloral has been given by Mr. Campbell, of
Dundee, in a case of puerperal convulsions, with such apparent
good effect, that I should recommend a further trial of it. The
case (b) was that of a primipara, aged 28, who was seized
with convulsions as soon as the labour pains made their
appearance. The convulsions recurred every half-hour, at
the commencement of each labour pain, and left the patient
perfectly unconscious during the intervals.

The patient looked rather ex-sanguine, with pulse weak.
and irregular, face, hands, and legs œdematous, and urine
slightly albuminous.

Five hours after the convulsions began, and while they were
still occurring at the same regular intervals, forty grains of
hydrate of chloral were given without any return of the
convulsions. Pains increased both in strength and frequency,
and she remained insensible and quiet, only being slightly
roused by the pains. Three hours later, while still asleep,

(a) *Vide* Author's paper on " Physical and Chemical Agents upon Blood
with special reference to the Mutual Action of Blood and the Respiratory
Gases." Phil. Trans., 1865, p. 713. Also in the Proc. of Roy. Soc., 1864, .
p. 159.

(b) *Lancet*, 30th July, 1870.

she was delivered of a dead child, and in three hours after delivery she was again seized with convulsions, which recurred, during six hours, with more or less severity every fifteen minutes, until thirty grains more of chloral were administered, when she again fell asleep, and woke next day, after a good night, without any further attack; both the œdema and albumen at the same time disappearing.

Tincture of iodine in albuminuria may by some be considered a novelty; but, nevertheless, it is a most successful remedy in cases of cerebral albuminuria, especially in those where the lesion or disease of the brain is not far advanced. In such, for example, as cerebral congestion arising from overwork, or excessive exposure to the sun, &c., I generally give two or three minims of the tincture along with ten of the aromatic spirits of ammonia; in the course of three weeks or so there is usually a marked diminution in the quantity of the albumen. A good illustration of this kind of treatment has recently been afforded me, in the person of one of our own profession. The gentleman, a provincial hospital surgeon, first consulted me in May, 1870. The urine was then of low specific gravity and loaded with albumen, which he attributed to gout; but which, on closely examining his symptoms, I suspected had, as well as the gout, a cerebral origin. Accordingly I put him on the tincture of iodine treatment, and saw no more of him till the month of July, when he returned, saying he was now also of my opinion, for within a fortnight from his first visit the albumen had begun gradually to disappear, and by a month had entirely vanished. At the same time, he said, he noticed that excitement, worry, or fatigue caused the albumen to reappear.

Iodine in combination with potassium has often been used with advantage in the treatment of kidney affections, and some have thought very large doses necessary. (a) M. Crocq,

(a) Crocq, *Lancet*, September 7th, 1867.

of Brussels, in the parenchymatous form of albuminous nephritis, commences with two or three grammes a day, increasing the dose by proportions of one gramme every two or three days, until he reaches the dose of ten or fifteen, and even twenty grammes. It is not always necessary to reach these high doses. Tolerance is easily obtained. He has never, he says, observed any inconvenience result from the administration of this remedy.

Astringent salts, in chronic albuminuria, as well as astringent acid tonics, are often of great service, especially in getting rid of the last traces of albumen after an attack of acute disease. Just as gonorrhœa often ends in a gleet, so it happens that the kidneys, after having for a long time been the seat of Bright's disease, continue to pour out small quantities of albumen long after all constitutional symptoms have disappeared. In such cases, sulphate of zinc, sesquichloride of iron, gallic acid, tannic acid, and the mineral acids have frequently a very salutary effect in checking the remnants of the albuminuria.

This action of the acids, at least, seems already to have found a physiological explanation, for Heynsius, while studying the diffusibility of albumen, (a) discovered that the exosmose of an albuminous to a saline solution is retarded by acidity, and accelerated by alkalinity; and, if such be the case out of the body, there can be little doubt, I think, that a similar, though perhaps a modified, effect follows the exudation of albumen from alkaline blood into acid urine.

True, doubts upon the value of iron in the treatment of renal disease have been thrown by Dr. Hassall, (b) finding that he obtained no ferruginous reaction from urine of patients to whom he had administered the remedy.

So long ago as 1852, while working in the laboratory of

(a) Douder's "Archiv.," vol. iv. part 3.
(b) *Lancet*, vol. ii p. 740, 1864.

Professors Wurtz and Verdeil, at Paris, I discovered that iron is a normal constituent of human urine, and published the fact in the *Pharmaceutical Journal* for November of that year in a paper entitled " Researches on the Colouring Principle of Urine, and the Existence of Iron in that Liquid." In a review of this paper in " Canstatt," 1853, Scherer denied the truth of the assertion that iron always existed as a normal constituent of human urine. Such a contradiction coming from a man of Professor Scherer's position, unjustly casting, as it did, a doubt either upon my powers of observation or veracity, so deeply wounded me that I at once started for Würzburg, and entered myself as a student in Scherer's own laboratory, determined to prove to him not only the existence of iron in healthy human urine, but even in the very "extractive colouring matter," which he himself had written so much about. This, I am happy to say, I did equally to his and my own satisfaction, as was shown in a paper which I read in the following spring before the Würzburg Medical and Physical Society, and published in its *Transactions*.(a) While working with Scherer I pushed the discovery a step farther, and found that iron actually exists normally in the liquid urine of all animals. After the salts of iron have been for a time administered medicinally, the quantity is increased in the urine ; and Dr. Hassall's inability to detect it arises, no doubt, from his testing the urine directly, instead of first evaporating and incinerating the residue, before applying the tests for the metal ; iron, like most other metallic substances, being quite undetectable when mixed with organic matter, unless present in an immense quantity.

As diet invariably plays an important part in all treatment, I must here call attention to the influence of foods on albuminuria, as previously pointed out in the physiological

(a) " Ueber Urohæmatin," &c. " Verhand der Phys. Med. Gesellschaft zu Würzburg," Bd. v., 1854.

part of our subject (p. 289). From what was then said, it will be readily understood why patients labouring under kidney disease should receive the lightest and most digestible kinds of diet. Moreover, their meals should be frequent and small rather than seldom and abundant.

The valuable researches of Professor Parkes (a) on the influence of food on the quantity of albumen eliminated by the kidneys, form the grounds for this latter recommendation. Dr. Parkes found that more albumen is eliminated after than before a meal, and that fasting not only invariably diminishes the quantity of albumen excreted, but even in some cases may cause it entirely to disappear from the urine—a fact which should make us avoid giving the patient more food than the wants of the system actually demand. The kind of food ought also to be regulated, in some measure, according to the form of kidney affection. Thus, while in cases of fatty renal degeneration, oleaginous diets are to be avoided, in the amyloid variety of the disease these same foods may be given, if not with advantage, at least with impunity. The general constitution and condition of the patient must, however, always be taken into consideration, and the diet selected according to the special requirements of the case.

As can be readily imagined pure albumen is an important adjunct to other treatment, for in order to keep the patient from losing strength by the constant drain of albumen from the blood, it is absolutely necessary that this substance be restored to the circulation through the medium of the food. In cases where the digestion is so weak that strongly albuminous foods, like beef-steaks, chops, and eggs, in the usual form are rebelled against, I simply give the white of fresh eggs, either raw or semi-cooked, by beating it up in hot water, beef-tea, milk or wine. I have not as yet met with

---

(a) *Med. Times and Gaz.*, 1854, p. 396.

a single case where the white of eggs has upset the digestive organs, although I have given the white of as many as six, eight, or even twelve eggs in the twenty-four hours. Where the patient tired of this albuminous food in one form, it was given in another, as above indicated, varying it as circumstances required.

Milk is another important agent in treatment. I do not now allude to skim milk, for of that I shall speak presently, but of good, fresh, wholesome milk. In the treatment of children it is indispensable, and even in the albuminuria of adults is of great service. It, like the egg, is Nature's own preparation of food. The milk may be given alone or made up into puddings, such as sago, tapioca, rice, macaroni, and suchlike; or in the more simple form of bread and milk, a sufficiency of sugar being added to render it palatable. It is necessary here to call attention to the fact that some stomachs rebel against cow's milk, and in these cases before abandoning milk diet it may be as well to try the effects of the milk of different kinds of animals, such as the ass, the mare, the goat, as these, especially in infants, are tolerated when it is found that woman's or cow's milk will not remain on the stomach. This observation has been specially suggested by the perusal of a fatal case of albuminuria in a child seven weeks old, recorded by (a) Mr. G. F. Helm, of Rugby, in which the affection (as that gentleman remarks) appeared to arise from the mal-assimilation of the albuminous elements of the food employed. This will be readily understood if we remember what was said at page 312, on the effect of imperfect digestion on the production of temporary albuminuria.

Skim Milk in Albuminuria.—In 1869, Dr. Donkin (b) pro-

_____

(a) *Lancet*, 18th January, 1869.　(b) *Ib.*, vol. ii. p. 768, 1869 ; *Ib.*, vol. i. p. 378, 1870 ; *Ib.*, vol. i, p. 603, 1871.

posed the plan of treating cases of Bright's disease, as well
as diabetes, by a purely skim milk diet. The amount
recommended to be taken being six or seven pints daily,
slightly warmed. He says it does not, however, do to begin
with this quantity. Two pints are sufficient at first, for
either in consequence of some peculiar idiosyncrasy, or
condition of the digestive system, a larger quantity may
produce sickness and diarrhœa.

Even later on, it is generally necessary to add something to
the milk to overcome the distaste which patients complain of
after a prolonged use of skim milk.

In the beginning of 1870, I found that a medical man at
Smyrna had advised a gentleman, about whom I was
consulted, to eat a raw onion along with the milk, and this
had the desired effect of overcoming his dislike to the milk
diet.

Notwithstanding my great partiality for milk diet in the
treatment of disease, I do not at all agree with those gentlemen who suppose skim milk to be a specific in any one form
of Bright's disease, diabetes (a) or disease of the supra-renal
capsules.

The skim milk recommended no doubt is an important
adjunct, as it presents us with an easily assimilated, non-
stimulating diet; one, too, containing a protein substance
casein, closely allied to the protein substance albumen,
which is being drained off from the blood by the kidneys,
and whose place must in some way or other be supplied in
order to prevent the patient dying from exhaustion. The
question, however, is, Is skim milk the best means of fulfilling
our object ?

In my opinion it has its advantages and disadvantages.

The former are, its being a non-stimulating, easily assimi-

(a) *Vide* letters in the *Lancet,* from Thorn, 19th February, also from
Donkin, May, and More, March, 1870.

lated food; the latter are its liquid form, and the great bulk necessary to be taken in order to obtain the requisite amount of solid material. I was once seriously asked by a barrister who was suffering from albuminuria if strawberries and cream were not a cure for Bright's disease; and on inquiring, was told that one of his friends had been "cured by eating them."

My reply was "certainly an admirable one, for it furnished the patient with a nourishing, non-stimulating diet, and allowed the healing powers of nature and other remedial agents time to do their work." I fear I look upon skim milk much in the same light as I did the strawberries and cream—an important adjunct to treatment, nothing more.

Before leaving the question of treatment, as a sequel to the general principles just enunciated, I may add a few words regarding the especial treatment of certain of the more common forms of albuminuria.

1st. In Acute Bright's Disease cupping, dry-cupping, leeching, or fomenting the loins ought immediately to be had recourse to, according to the urgency of the symptoms and the constitution of the patient. In the beginning, diuretics must be scrupulously avoided, and even in the second stage only those of the antiphlogistic variety are admissible. The bowels are to be made to act freely; the action of the skin and lungs increased. The quickest, and in many cases the safest, way of doing this is by using the hot-air bath as described at page 334; and it is astonishing how rapidly beneficial its employment may sometimes be. On one occasion I was very forcibly struck with this, in the case of a little child, who speedily recovered, after being almost moribund from the united effects of dropsical effusion and uræmic intoxication. The case was one that I saw along with Mr. Brown, of Finsbury-circus, and, if I remember rightly, the age of the child was not more than fifteen or eighteen months.

2nd. In Fatty Kidney, in addition to the means already indicated, great attention must be paid to the improvement of the general health, for in the majority of these cases the affection is more frequently due to constitutional peculiarity than to an accidental cause. Fresh air and regular exercise within the margin of fatigue, together with light, non-oleaginous foods, are to be prescribed. Change of air and hot salt-water baths ought, in the early stages of the affection, alike to be had recourse to. When tonics are given they should be of the astringent and acid kind, except when counter-indicated by any constitutional peculiarity.

3rd. In the Amyloid variety of Renal Degeneration, which, as already shown, is frequently the result of a syphilitic taint, the judicious employment of the bichloride of mercury is often of much use; but care must be taken to stop as soon as the mercurial fœtor of the breath becomes distinct, and before the gums are affected, for such patients do not readily tolerate the reducing action of mercury. The calomel vapour bath is likewise sometimes very useful. It will be often necessary to combine with this treatment the administration of port wine and light nutritious foods.

4th. Atrophied Kidney must be treated upon the general principles already enunciated, and, of course, it is one of the most suitable of all renal affections for increasing the action of the skin and lungs. Milk diet is here of the utmost value, and so also the pure albumen treatment described at page 343.

In it, as well as in all cases of confirmed albuminuria, we may occasionally be enabled to prolong the life of our patient from the free action of the skin being periodically induced by the Turkish bath, and when that cannot be obtained the hot vapour bath may often be substituted with advantage, care being always taken to employ the bath periodically and at short intervals; for then alone can we expect advantageously to make the skin vicariously perform the function of the renal organs.

In the latter stages of atrophied kidney, as well as in the fatty and waxy forms of kidney disease, the symptoms of uræmic poisoning may sometimes be diminished by administering small doses of tartar emetic. Indeed, Dr. Lang, of Königsberg, recommends the use of tartarised antimony in the uræmia that follows acute exanthematous diseases. (a) My own experience leads me to prefer effervescing saline mixtures. Small doses of Liquor Strychniæ and Vini Ipecacuanhæ are frequently of much service.

5th. Regarding Puerperal Albuminuria, sufficient has already been said at page 338 to serve as a guide to its treatment. All that need here be added is, that premature labour ought only to be had recourse to in those forms where the pregnancy is associated with true kidney disease, or where there is reason to anticipate the occurrence of convulsions either at or before the time of delivery. Premature labour may also successfully be had recourse to in the earlier months of pregnancy, when the health of the mother is rendered precarious by the constitutional disturbances arising from the albuminuria.

Thus, Dr. Lee related a case to the Medical-Chirurgical Society in 1863, in which he had induced the expulsion of a fourth months' fœtus in a pregnancy complicated with albuminous urine, dropsy, and amaurosis, after which the albumen gradually disappeared, and the vision of the patient improved.

When in the course of puerperal albuminuria convulsions have once manifested themselves, the readiest and safest means of subduing them appears to be by the free administration of an anæsthetic, as pointed out at page 339.

(a) *Brit. Med. Jour.*, January 21, 1859.

6th. As regards Albuminuria associated with Organic Disease of the Heart, Liver, or Pancreas, as well as in that variety the concomitant of general affections, the principles already indicated must serve as the landmarks to the treatment of the renal disturbance; whereas the other disease or diseases forming the complication are to be treated according to their special requirements; always remembering that the removal of the cause, rather than the mitigation of its effects, ought to be the first consideration.

7th. Albuminuria with Dropsy. In addition to the general principles previously enunciated, it may be as well to remark that when the dropsy becomes, as it frequently does, one of the most distressing symptoms to the patient, as well as the most baffling to the skill of the medical attendant, totally defying the action of elaterium, jalap, and other remedies.

Immense benefits often follow the mechanical removal of the fluid from the œdematous limbs by acupuncture. Many practitioners object to the employment of acupuncture in dropsical patients, on the grounds that its beneficial effects are only transitory, and because some sloughing occasionally follows at the seat of the wounds. My experience, which has now been pretty extensive, leads me to the very opposite conclusion. In the first place I have never seen sloughing follow the use of the needle; whereas, on the other hand, I have witnessed very extensive sloughing in dropsical patients where no such remedial means have been had recourse to. The very worst example of this kind was in the case of a lady, whom I saw in consultation along with Mr. Marshall, of Bedford-square, where over both the lower limbs were ugly black sloughs varying in size from a shilling to that of the palm of the hand, which a timely use of the needle would, in my opinion,(a) have probably prevented.

(a) I prefer the tubular steel needle to anything else.

Again, so far from the benefit of acupuncture being only transitory, I have known, when properly done, its beneficial effects kept up, not only for days, but for weeks. As a good example of the value of artificial draining in cases of cardiac albuminuria with extensive anasarca, I cannot do better than allude to a case I several times saw in consultation with Dr. J. M. Ramsbotham. From his ably written report is taken the following extract, which is all the more valuable as a contribution to the question of treatment by acupuncture, from the fact of the writer being totally unaware of the point which the report was intended to illustrate. Consequently Dr. Ramsbotham has given no special prominence to the question at issue, but simply related *inter alia* the effects of the artificial drainage by acupuncture.

The patient, a gentleman aged 37, suffering from mitral disease with hypertrophy and dilatation of right ventricle, was first seen by us in January, 1869. In February the urine was ascertained to be of a s. g. of 1015, and loaded with albumen. Under treatment the albumen diminished and the patient gradually improved, until, contrary to medical advice, he went on a visit to Hastings. Here I take up Dr. Ramsbotham's own report, and give it verbatim :—

" In the beginning of June, 1869, after a visit to Hastings without sanction, matters assumed a very serious aspect, all the symptoms being greatly aggravated, orthopnœa very distressing, the only mode of getting even a short nap was by resting the forehead on a pillow, fixed against the bed-post, secretion of urine diminished, with increased amount of albumen, anasarcous condition of the whole body, face, arms, chest, abdomen, the penis and scrotum enormously distended, thighs and legs more than twice the natural size, brown tongue, rambling, and threatening uræmia ; indeed, the condition was such as to warrant the opinion that he would not live forty-eight hours. The treatment consisted of enclosing the legs in a hot-air bath, after puncturing them freely; the administration of Tinct. Digitalis and Scillæ in Decoct. Scoparii. The amount of fluid discharged from the punctures was enormous, and in some parts of the leg threatened to slough, but the good effect of the drainage was very decided; the swelling in different parts of the body began to subside and gradually recovered the normal size, with the exception of the thighs and legs, which always remained, notwithstanding a moderate amount of constant drainage. By the 14th of July he had so far recovered as to be able to call upon Dr. George Harley, and continued so to improve up to the middle of August, that is to say, there was less dyspnœa and embarrassment of heart's action, less albumen, but the urine never free from it; the legs continued still much enlarged, with great thickening of the integuments, but still discharging freely and constantly. At the end of August he had a relapse without any apparent cause, aggravation of all the symptoms, but not to the alarming extent of the attack in June--this attack also subsided as before, and under much the same treatment. He was seen again by Dr. George Harley about the end of November, and continued much in the same condition till the end of January, 1870, at which time there appeared to be a great improvement or rather abeyance, that is to say, less dyspnœa and cardiac symptoms, more sleep, better appetite, more action of skin and kidneys, but the urine still contained a large amount of albumen, the thighs and legs continued very much enlarged, but discharging freely; and he was enabled occasionally to leave his bed-room and get down-stairs, but as there was some difficulty in re-mounting, this luxury was not often indulged in.

"Mr. F. continued much in the same condition, with but little variation, till the middle of May, when it was his opinion that he should never get better unless he went out of doors (notwithstanding all the plain speaking he till then clung to the idea that he should get better) ; this he accordingly did, having a drive in a carriage, but this not being sufficient to satisfy him, he determined to call at his father's house in the same street, four doors from his own, discharging the carriage, further determining to walk home. This feat was accomplished, but with great difficulty, followed by increased dyspnœa, and great aggravation of all his symptoms. On the 24th May he had suppression of the urine, and the drain from the legs ceased ; he became uræmic, and died on the 28th of the same month."

This case requires no comment from me. It speaks for itself, seeing that the artificial drain from the legs was beneficially kept up from the beginning of June, 1869, until the end of May, 1870, a few days less than a year.

I am glad to find Dr. Handfield Jones, in a communication made to the Clinical Society, 10th February, 1871, advocating this line of treatment. He treated a case of anasarca by making a single puncture in the calves of both legs with a fine trocar, and, after withdrawing the stilette, left the canula open for several hours, to allow the fluid to drain away. In this manner he succeeded in the first operation in drawing off sixty measured ounces of fluid from the right leg, but only ten from the left, in consequence, he supposed, of the canula not lying properly in the subcutaneous cellular tissue. In a second operation on the same man, three days afterwards, Dr. Handfield Jones drew off 120 ounces of fluid, besides a great deal which ran from the punctures for several days afterwards, sufficient to saturate three blankets. For the performance of the operation, the man was placed in a sitting posture, and this he considered important, as it facilitated the draining away of the fluid. (a) In the course of the discussion which followed the reading of the paper, Dr. Hilton Fagge judiciously

---

(a) *Lancet*, 4th March, 1871.

remarked that the surrounding skin should be thoroughly greased, so as to avoid all contact between it and the discharge.

A great deal more might be written regarding Albuminuria; but I think that enough has been said for my present purpose, which has been merely to point out the general principles which ought to guide our diagnosis and treatment.

Every one must feel convinced that the employment of the microscope is as indispensable to treatment as to diagnosis. Thus, Aitken says:—

" Without the microscopic examination of the urine from day to day it is impossible to distinguish between a case likely to improve under treatment, and one which may be viewed as hopeless, and without the daily use of the microscope the treatment becomes at the best but merely guess-work."(a)

It must be borne in mind, too, that as one swallow does not make a summer, neither does one tube-cast in kidney disease at any time suffice to establish an exact diagnosis. This can be easily understood when it is recollected that the deposits in the urine are but the representatives and results of the morbid change occurring in the renal tissue at the moment of their detachment; and as disease is never stationary, it is only by comparing the character of the renal *débris* of one period with that of another that we are enabled to establish the exact character of the affection, and discover whether it is marching on to a fatal termination or gradually retrograding to a point within the boundaries of healthy action.

(a) Aitken's "Science and Practice of Medicine." 2nd ed., vol. ii. p. 324.

INDEX.

The references to the subjects of entire chapters and their sections are printed in *italics*.

368 . INDEX.

Pardon and Son, Printers, Paternoster Row, London.